THE McGRAW-HILL CIVIL ENGINEERING PE EXAM DEPTH GUIDE
Geotechnical Engineering

THE McGRAW-HILL CIVIL ENGINEERING PE EXAM DEPTH GUIDE

Geotechnical Engineering

S. Joseph Spigolon, Ph.D., PE

McGRAW-HILL
New York   Chicago   San Francisco   Lisbon   London   Madrid
Mexico City   Milan   New Delhi   San Juan   Seoul
Singapore   Sydney   Toronto

Cataloging-in-Publication Data is on file with the Library of Congress

McGraw-Hill
A Division of The McGraw·Hill Companies

Copyright © 2001 by The McGraw-Hill Companies, Inc. All rights reserved. Printed in the United States of America. Except as permitted under the United States Copyright Act of 1976, no part of this publication may be reproduced or distributed in any form or by any means, or stored in a data base or retrieval system, without the prior written permission of the publisher.

1 2 3 4 5 6 7 8 9 0 AGM/AGM 0 7 6 5 4 3 2 1

ISBN 0-07-136184-7

The sponsoring editor for this book was Larry S. Hager and the production supervisor was Sherri Souffrance. It was set in Times Roman by Lone Wolf Enterprises, Ltd.

Printed and bound by Quebecor/Martinsburg.

 This book is printed on recycled, acid-free paper containing a minimum of 50% recycled, de-inked fiber.

McGraw-Hill books are available at special quantity discounts to use as premiums and sales promotions, or for use in corporate training programs. For more information, please write to the Director of Special Sales, McGraw-Hill, Professional Publishing, Two Penn Plaza, New York, NY 10121-2298. Or contact your local bookstore.

Information contained in this work has been obtained by The McGraw-Hill Companies, Inc. ("McGraw-Hill") from sources believed to be reliable. However, neither McGraw-Hill nor its authors guarantee the accuracy or completeness of any information published herein, and neither McGraw-Hill nor its authors shall be responsible for any errors, omissions, or damages arising out of use of this information. This work is published with the understanding that McGraw-Hill and its authors are supplying information but are not attempting to render engineering or other professional services. If such services are required, the assistance of an appropriate professional should be sought.

CONTENTS

Preface xii
About the Author xiii

CHAPTER 1: WHAT IS GEOTECHNICAL ENGINEERING? 1.1
Scope of This Quick-Study Guide 1.2

CHAPTER 2: IDENTIFICATION AND CLASSIFICATION OF SOIL AND ROCK 2.1
Definition of Soil and Rock 2.1
Soil Identification 2.1
    Grain Size Distribution 2.1
    Atterberg Limits 2.4
Soil Classification Systems 2.6
    Unified Soil Classification System 2.7
    AASHTO Soil Classification System 2.10
    USDA Soil Taxonomy 2.10
Rock Identification and Classification 2.12
    Intact Character 2.12
    In Situ Character 2.12
Weight-Volume (Phase) Relationships 2.15
Example Problems 2.16
    Grain Size Distribution 2.16
    Atterberg Limits 2.18
    Soil Classification 2.18
    Rock Description and Classification 2.20
    Weight-Volume (phase) Relationships 2.22

CHAPTER 3: HYDRAULIC PROPERTIES OF SOIL AND ROCK 3.1
Coefficient of Permeability 3.1
    Laboratory Tests for the Coefficient of Permeability 3.3
    Equivalent Permeability of Stratified Deposits 3.4

        Empirical Estimates for Coefficient of Permeability 3.6
        Permeability of Rock 3.7
        Field Tests for Permeability by Pumping from Wells 3.8
Effective Stress and Seepage Pressure 3.11
Principle of Effective Stress 3.12
        Seepage Pressures and the Critical Hydraulic Gradient 3.13
Seepage of Water Through Soils 3.15
        Flow Net in Isotropic Soil 3.15
        Drawing a Flow Net 3.17
        Flow Net in Anisotropic Soil 3.17
        Seepage Line—Phreatic surface 3.18
        Heaving of Soil at Exit Point 3.19
Example Problems 3.20
        Constant Head Permeability Test 3.20
        Falling Head Permeability test 3.20
        Equivalent Horizontal Permeability 3.20
        Equivalent Vertical Permeability 3.21
        Estimating the Coefficient of Permeability 3.21
        Well Pumping Test—Fully Penetrating Artesian Flow 3.21
        Well Pumping Test—Partially Penetrating Artesian Flow 3.22
        Well Pumping Test—Fully Penetrating Gravity Flow 3.22
        Well Pumping Test—Partially Penetrating Gravity Flow 3.23
        Effective Stress 3.23
        Seepage Pressures and Critical Hydraulic Gradient 3.24
        Flow Net in Isotropic Soil 3.25
        Heaving of Soil at Exit Point 3.26

CHAPTER 4: COMPRESSIBILITY OF SOIL AND ROCK   4.1

Compressibility 4.1
        One-Dimensional Compression 4.2
        Laboratory Oedometer Test 4.4
        Determining Preconsolidation Pressure 4.5
        Determing the field void Ratio-Pressure Relationship 4.7
        Calculation of Settlement Due to One-Dimentional Primary Consolidation 4.9
Vertical Stresses Due to Surface Loads 4.10
        Boussinesq Equations 4.10
        Stresses Under Rigid Footings 4.15
Consolidation Settlement of Structures 4.18

Foundation Settlement in Cohesionless Soil 4.18
Time Rate of Consolidation 4.18
    Theoretical development 4.18
    Laboratory Consolidation Test 4.20
Example Problems 4.24
    One-Dimensional Compression 4.24
    Vertical Stresses Due to Surface Loads 4.25
    Foundation Settlement in Cohesive Soil 4.28
    Time Rate of Consolidation 4.30

CHAPTER 5: STRENGTH OF SOIL AND ROCK   5.1

Combined Stresses 5.1
Mohr's Circle of Stress 5.2
Mohr-Coulomb Failure Criteria 5.3
Tests for Shear Strength of Soils 5.5
    Direct Shear Test 5.5
    Traxial Shear Test 5.5
    Types of Laboratory Shear Tests 5.6
    Pore-Water Pressure Resulting from Stress 5.8
    Relative Density 5.9
    Definitions 5.9
    Laboratory Relative Density Tests 5.9
Effect of Initial Relative Density on Shear Strength 5.10
    Negative Pore-Water Pressure Due to Partial Saturation 5.11
Shear Strength Relationships 5.12
    Consolidated-Drained (S-Test) Shear of Saturated Soils 5.12
    Consolidated-Undrained (R-Test) of Saturated Soils 5.15
    Unconsolidated-Undrained (Q-Test) Shear of Saturated Soils 5.16
    Shear Strength of Partially Saturated Soils 5.17
    Sensitivity of Clays to Disturbance 5.18
    Selection of Shear Strength for Design 5.19
Example Problems 5.20

CHAPTER 6: ENGINEERING GEOLOGY OF ROCKS AND SOILS   6.1

Origin of Rocks and Soils 6.1
    Igneous Rocks 6.1
    Sedimentary Rocks 6.1

    Metamorphic Rocks 6.1
Soil Forming Processes 6.2
    Weathering Processes 6.2
Natural Soil Deposits 6.4
    Residual Soils 6.4
    Transported Soils 6.5
Soil Structures 6.11
    Cohesionless Soil 6.11
    Cohesive Soil 6.11
    Mixed Grain Soil 6.12
Example Problems 6.13

CHAPTER 7: ENGINEERING SUBSURFACE INVESTIGATIONS   7.1

Plan for a Subsurface Investigation 7.1
Sources for Pre-Existing Information 7.2
    Geologic Data Sources 7.2
    Project Records 7.3
    Remote Imaging 7.3
    General Sources 7.3
Engineering Geophysical Methods 7.4
    Electrical Resistivity Surveys 7.4
    Seismic Refraction Surveys 7.4
Subsurface Access Methods 7.6
    Pits and Trenches 7.6
    Borings 7.7
Sampling of Soils and Rocks 7.10
    Thick-Wall Split-Barrel Drive Samplers 7.10
    Vibrating tube Sampler 7.11
    Thin-Wall Tube Samplers 7.11
    Diamond-Core Sampler 7.13
In Situ Testing of Soils and Rocks 7.14
    Plate Load Test 7.14
    Field Vane Shear Test 7.16
    Borehole Shear Test 7.16
    Unconfined Compression test of Undisturbed Cohesive Sample 7.17
    Handheld Devices: Penetrometer and/or Torvane Test of Cohesive Sample 7.18
    Standard Penetration Test 7.19
    Dynamic Penetrometer Test, Thick-Wall Tube or Solid Cone 7.21
    Handheld Sounding Rod Test 7.23

Analysis of Penetration Test Data 7.23
    Estimating Unconfined Compressive Strength from SPT Data 7.23
    Estimating Relative Density and Friction Angle from SPT Data 7.24
    Estimating Unconfined Compressive Strength from CPT Data 7.25
    Estimating Pre-consolidation Pressure 7.27
    Estimation of Liquefaction Potential 7.29
Planning for a Subsurface Investigation 7.32
    Factors Affecting the Plan for an Investigation 7.32
    How Many Borings, Where, and How Deep? 7.32
Example Problems 7.36

CHAPTER 8: SHALLOW FOUNDATIONS—FOOTINGS AND RAFTS   8.1

Bearing Capacity 8.1
Settlement 8.1
Shear Failure Modes 8.2
    General Shear Failure 8.2
    Punching Shear Failure 8.2
    Local Shear Failure 8.2
Theoretical Bearing Capacity Equations 8.3
    Footings of Finite Length 8.4
    Footings with Inclined Load 8.4
    Shallow Footings on or Near Slope 8.4
    Bearing Capacity of Cohesive Soils 8.8
    Settlement Limitations 8.8
Bearing Capacity from In Situ Tests 8.11
    Footings on Sand and Nonplastic Silt 8.11
    Rafts on Sand and Nonplastic Silt 8.12
    Footings on Clay and Plastic Silt 8.13
    Rafts on Clay and Plastic Silt 8.13
    Earthquake Loading of Foundations 8.14
    Modulus of Subgrade Reaction 8.14
Example Problems 8.16

CHAPTER 9: DEEP FOUNDATIONS—PILES AND PIERS   9.1

Types of Deep Foundations 9.1
Deep Foundation Failure Modes 9.2
Deep Foundation Capacity from Load Tests 9.3

Field Loading Tests in Axial Compression 9.4
Tension and Lateral Loading Tests 9.4
Test Loading and Time Relationship 9.5
Interpretation of Pile Load Tests 9.5
Deep Foundation Capacity from Static Analysis 9.7
Load Carrying Capacity of a Single Pile or Pier in Granular Soil 9.8
Load Carrying Capacity of a Single Pile or Pier in Cohesive Soil 9.10
Lateral Loading of Piles and Piers 9.12
Bearing Capacity of Pile Groups 9.12
Settlement of Pile Groups 9.13
Pile Capacity from Driving Data 9.14
Engineering News Formula 9.14
Danish Formula 9.14
Use of Pile Driving Formulae 9.14
Example Problems 9.15

CHAPTER 10: RETAINING STRUCTURES 10.1

Types of Retaining Structures 10.1
Gravity Retaining Walls 10.1
Structural Retaining Walls 10.2
Flexible Retaining Walls 10.2
Lateral Earth Pressure 10.2
Active Earth Pressure 10.3
Rankine's Theory 10.5
Cohesive Soils 10.6
Surcharge Loading 10.6
Active Lateral Pressure Due to Earthquake Forces 10.7
Passive Earth Pressure 10.8
Design of Rigid Retaining Walls 10.9
Design of MSE Retaining Walls 10.10
Design of Flexible Retaining Walls 10.12
Design of Braced Excavations 10.18
Example Problems 10.19

CHAPTER 11: SLOPE STABILITY 11.1

Types of Slope Failures 11.1
Sliding Wedge Analysis 11.2

Infinite Slope Analysis 11.5
Slope Stability Charts for Homogeneous Slopes 11.6
Method of Slices 11.8
Seismic Slope Stability Analysis 11.10
Example Problems 11.10

CHAPTER 12: COMPACTED FILL 12.1

Purpose of Compaction 12.1
Basic Concepts 12.2
    Zero Air Voids Curve 12.2
    Pressure Distribution with Depth 12.2
Compaction of Cohesionless Soils 12.2
    Factors Affecting Field Compaction of Cohesionless Soils 12.3
    Effect of Soil Gradation 12.3
    Effect of Water Content 12.4
    Effect of Lift Thickness and Magnitude of Effort 12.4
    Effect of Characteristics of the Compaction Equipment 12.5
Compaction of Cohesive Soils 12.6
    Example of a Test Fill 12.7
    Effect of Soil Texture 12.8
    Effect of Lift Thickness 12.9
    Effect of Field Compaction Equipment 12.9
    Effect of Compacting Moisture on Engineering Properties 12.9
    Modifying Compactibility with Chemicals 12.12
Quality Assurance Testing 12.12
    Quality Assurance Definitions 12.12
    The Proctor Compaction Test for Cohesive Soil 12.13
    Field Density and Water Content Test Methods 12.14
    Laboratory and Field California Bearing Ratio Test 12.16
Specifying Soil Compaction 12.17
    Acceptable Soil Types 12.17
    Compactive Effort Requirement 12.18
    Types of Compaction Equipment 12.18
    Lift Thickness 12.19
Example Problems 12.20

INDEX I.1

PREFACE

This book was written as a quick-study reference guide to the main topics of geotechnical engineering for specialists in that field. It certainly is not intended to replace any of the many fine textbooks on the subject. Concepts and equations, both theoretical and empirical, are presented but are not fully developed. There is no discussion of the structural design of substructure elements. As an aid in studying for the professional engineer exam, only those analysis topics that are most likely to appear on the exam are covered. For example, there are no computer programs since it is highly unlikely that they will be included in an examination.

There are three fundamental requirements for a complete analysis used in a geotechnical engineering design, and this book deals with second and third items in this group:

1. Must know the full loading conditions imposed by the structure to be supported
2. Must know the applicable engineering properties of the material (soil or rock) to be used to support the structural loads
3. Must use the appropriate analytic model for evaluating the interaction between the structural loads and the configuration and properties of the supporting soil or rock and must know and understand the premise and limitations of the chosen model

Engineering is about design, and geotechnical engineering is about design for construction involving soil and rock as a structural medium. It is assumed that the reader has completed all of the normal undergraduate courses in an ABET accredited civil engineering program, including mathematics, the basic sciences, the engineering sciences, and design courses in several of the areas of civil engineering. This sequence of courses prepares the fledgling engineer with the necessary design philosophy and procedures for any aspect of civil engineering. It is also assumed the reader has had at least four years of professional geotechnical engineering design experience, under the guidance of a qualified geotechnical engineer or the equivalent in advanced study prior to the examination for the professional engineering license.

This is not a field for persons with only limited knowledge of the subject. As with all other fields of civil engineering, many of the theoretical and empirical analysis procedures have been simplified to make the calculations easier and quicker for the qualified practitioner. Simply filling in the blanks in a computer program or in what is seemingly a simple procedure or equation without knowing the premise or the limitations of the program, procedure, or equation can lead to physical and/or economic failure.

The geotechnical engineer must understand physical geology to the extent that he/she can review published geologic literature and geologic maps and develop a sense for the probable near-surface soil profile before planning the subsurface investigation. The field and laboratory testing program can only be effectively planned and executed if the engineer fully understands the requirements of the applicable theoretical and empirical models and the advantages and limitations of field and laboratory tests. Given a good estimate of the near-surface soil-rock profile and the engineering properties of the materials encountered, it is necessary to use the appropriate analytic model. Only then can an effective geotechnical engineering design be made.

S. Joseph Spigolon, Ph.D., PE

ABOUT THE AUTHOR

S. JOSEPH SPIGOLON, Ph.D., PE

S. Joseph Spigolon, Ph.D., PE, is a geotechnical and materials engineer with 45 years' experience, including 15 years of university teaching and 30 years of consulting. He is the author of many published articles.

CHAPTER 1
WHAT IS GEOTECHNICAL ENGINEERING?

Soils and rocks are the subject of several scientific and professional disciplines, including agronomy, soil science, geology, geophysics, engineering geology, geological engineering, mining engineering, and geotechnical engineering. How do these disciplines differ from each other; that is, what are their primary concerns about soils and rocks?

Agronomists and soil scientists are interested primarily in those uppermost few feet of the earth's surface that support plant life. This zone contains the highly weathered organic and mineral soils. The underlying parent material is of interest chiefly as a source of the overlying pedologic soil.

Geologists study the earth's present and past morphology and structure, its environments, and the fossil record of its inhabitants. As scientists, they study without regard to the specific application of their knowledge. They use inductive reasoning to reach a general conclusion from their observations. Their principal concern is with rock formations at and well below the earth's surface.

Geophysicists use active and passive energy (geophysical) methods to identify geologic features, determine some of their engineering properties, and detect hidden cultural features.

Engineering geologists investigate and interpret the near-surface geologic profile so they can identify potential geologic hazards and provide a qualitative model of the subsurface, primarily rock deposits, at specific construction sites. Their findings are used by geotechnical and structural engineers in the design of a structure or facility.

Geological and mining engineers are concerned with the exploration for, and commercial development of, mineral and fuel deposits and groundwater resources.

Civil engineers are concerned primarily with the design of structures and facilities for the benefit of the public. Their work involves the quantitative analysis of economically feasible alternatives for, the preparation of plans and specifications for, and the monitoring of construction of public structures. All engineered structures and facilities are ultimately supported on or in the near-surface soil and rock.

Geotechnical engineers are civil engineers who specialize in using soil and rock as a structure-supporting medium. They apply their special knowledge and experiences of soil and rock behavior, combined with the principles of soil mechanics, rock mechanics, structural analysis, hydraulics, and geological science, to the quantitative engineering design of foundations, retaining structures, landslide stabilization works, and earthworks. Geotechnical engineering includes, but is not limited to:

1. *Subsurface investigations* involve the planning, specification, and evaluation of field and laboratory investigations of the *in situ* stratification and engineering properties of soil and rock deposits and of the groundwater regime. They are performed to provide appropriate supporting data for use in the preparation of quantitative engineering design recommendations reports.
2. *Geotechnical engineering* design involves analyzing data from subsurface investigations, making design calculations by the selection of suitable soil and rock properties and their application in appropriate analytic models. The design also involves the preparation of quantitative engineering design recommendation reports and/or plans and specifications, and monitoring the construction of those aspects of civil engineering works involving soil and rock such as
    a. Shallow and deep foundations for the support of structures, including foundations for structures involving vibrating, pulsating, or impacting machines and devices and the foundations for structures at sites subject to earthquake-induced hazards such as severe shaking, liquefaction, sliding, and tsunami inundation
    b. Flexible and rigid retaining structures such as walls, excavation sheeting, and bracing, quays, cofferdams, and abutments
    c. Earthworks such as earth- and rock-fill dams, embankments, levees, and dikes
    d. Compacted soil and rock fills, including the quantitative evaluation of the stability of cuts in natural soil and rock and of the slopes of man made fills
    e. Landslide mitigation and repairs
    f. Slope stabilization methods and systems
    g. Embankments, subgrades, and bases for road and airfield pavements
    h. Improvements in the structural quality of soil and rock masses by methods such as precompression, dynamic compaction, drainage blankets, wick drains, geosynthetics, chemical additives, and grout
    i. Site drainage and erosion control materials and methods, including the investigation and quantitative analysis of soil and rock permeability and groundwater flow patterns as they affect the stability of structures and earthworks.

SCOPE OF THIS QUICK-STUDY GUIDE

The remainder of this book is presented in five major sections corresponding generally to the topics listed above. The major section headings are (a) Soil Mechanics, (b) Subsurface Investigations, (c) Foundations, (d) Retaining Structures, and (e) Earthworks.

The Soil Mechanics chapters deal with the identification and classification of soils and rock and with the theoretical aspects of their hydraulic properties, compressibility, and shear strength.

The *Subsurface* Investigations chapters present the engineering geology of soils and rocks and geotechnical engineering subsurface investigations.

The *Foundations* chapters present the engineering of shallow footing and raft foundations and of deep pile and pier foundations.

The *retaining Structures* chapters discuss the design of excavation and bracing and of rigid and flexible retaining walls.

Finally, the *Earthwork* chapters deal with analytic methods for slope stability and discuss the various aspects of soil compaction.

Example problems are included at the end of each chapter, complete with solutions, covering the topics surveyed in that chapter. Each problem has been selected carefully to illustrate fully the applications of one or more of the theoretical or empirical concepts and/or analytic methods that were introduced in the chapter.

CHAPTER 2
IDENTIFICATION AND CLASSIFICATION OF SOIL AND ROCK

DEFINITION OF SOIL AND ROCK

Civil engineers distinguish between two kinds of natural, structure-supporting or structure-forming media—rocks and soils. Rocks are natural aggregations of mineral grains connected by permanent and strong cohesive bonds. Soils are natural aggregations of mineral grains that are the weathering product of rocks and consist of discrete grains that have been formed by mechanical or chemical-biological breakdown. The difference between rock and soil is not always clear, and they are often distinguished by the ease with which particles can be separated by mechanical means, such as by a hand-held pick or by earth moving equipment.

SOIL IDENTIFICATION

Soil identification is based on factual data—field and laboratory soil tests and observations—used to support a soil description. The two identification tests of soil materials most useful as indicators of potential engineering behavior, or *index properties*, are the grain size distribution test and the Atterberg limits test.

Grain Size Distribution

Sieve (mechanical) Analysis Test. A sample of soil is dried and then passed (shaken) through a series of screens having square openings of specific dimensions, each successive screen having smaller openings than the previous one. The weight of the soil retained on each screen is accumulated so that the percentage of the total that is finer or coarser than each screen size can be tabulated or graphed, as shown in Figure 2.1.

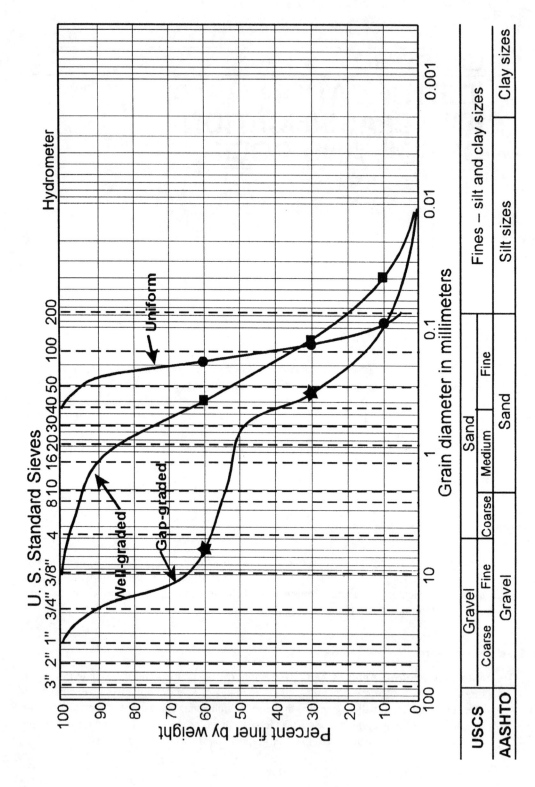

FIGURE 2.1 Typical grain size distribution curves.

The finest practical screen size is the No. 200, which has 200 openings per lineal inch (2.54 cm) in both directions. Finer screens (more openings per inch) tend to clog during the shaking test and, therefore, are not used. Soil descriptions based on grain size are shown in Table 2.1.

Hydrometer test. This test is used for analysis of the distribution of soil grain sizes that pass through the No. 200 sieve. A weighed sample of the soil is placed in a water column and allowed to settle quietly in the water. At successive times, a hydrometer placed just below the surface of the water column measures the density of the soil-water slurry remaining near the surface. The rate at which the soil particles settle, measured by the change in slurry density at hydrometer level, is evaluated by using Stokes' law for the rate of sedimentation of particles in a fluid, permitting calculation of the amounts of each of the equivalent spherical diameters (grain sizes) in the sample.

The mechanical weathering of rock tends to create a statistical distribution of particle size groups whose frequency histogram can be approximated by a Gaussian (normal), or bell-shaped, curve. This appears to be valid even for uniform soils that have been well sorted by a transportation process such as water or wind. The cumulative distribution function of a normal curve forms an S-shaped (or reverse S-shaped) curve on an arithmetic scale. The semi-logarithmic graph of Figure 2.1 is used because, although the percent finer scale is based on weight that is proportional to volume, the grain size scale is based on the equivalent spherical diameter of the particles that is proportional to the cube root of the volume. A power curve becomes an arithmetic function when plotted as a logarithm.

As shown in Figure 2.1, the distribution of grain sizes can be (a) *well-graded*, having an assortment of sizes from the largest to the smallest; (b) *uniform,* being nearly all one size; or (c) *gap-graded*, having one grain size fraction missing, or being a combination of

TABLE 2.1 Soil Description Based on Grain Size [1].

Soil Name	Diameter		U. S. Standard Sieve Size	Familiar Example
	mm	inches		
Boulders	Over 300	Over 12	> 12"	Larger than basketball
Cobbles (rounded)	76-300	3–12	3"-12"	Grapefruit
Coarse Gravel	19-76	0.75-3	0.75"-3"	Orange or lemon
Fine Gravel	4.75-19	0.19-0.75	No. 4-0.75"	Grape or pea
Coarse Sand	2.0-4.75	0.08-0.19	No. 10-No. 4	Rock salt
Medium Sand	0.42-2.0	0.016-0.08	No. 40-No. 10	Sugar, table salt
Fine Sand	0.074-0.42	0.003-0.016	No. 200-No. 40	Powdered sugar
Silt sizes	0.002-0.074	0.00008-0.003	Rock flour and finer; particles cannot be distinguished with naked eye at a distance of 20 cm (8 inches)	
Clay sizes	Less than 0.002	< 0.00008		

two soils with different grain-size distributions. Two coefficients are used to describe the configuration of the grain size distribution curve:

1. The coefficient of uniformity, $C_u = \dfrac{D_{60}}{D_{10}}$ (2.1)

2. The coefficient of curvature, $C_z = \dfrac{D_{30}}{D_{60} \times D_{10}}$ (2.2)

where D_{60}, D_{30}, and D_{10} are, respectively, the grain diameters corresponding to 60 percent, 30 percent, and 10 percent finer. For a well-graded gravel, the coefficient of uniformity C_u is greater than 4. For a well-graded sand, the coefficient of uniformity C_u is greater than 6. For both gravel and sand, the coefficient of curvature C_z must fit within the range $1 < C_z < 3$.

Atterberg Limits

The relationship between the consistency and water content of a remolded clayey (cohesive) soil was defined in 1911 by A. Atterberg, an agronomist who was concerned with the tillability of agricultural soil [2]. Atterberg developed a procedure to define the transitions between the various consistency states that a remolded soil goes through as its water content changes. Atterberg's procedure has been refined and now is referred to as *ASTM Standard Method D 4318* [3]. The amounts of water that correspond to the boundaries between the consistency states are called the *Atterberg limits*.

Referring to Figure 2.2, consider a remolded clayey soil whose water content is so high that it is essentially a soil-water slurry. As the water content is reduced by slow-drying, the soil mass passes from a liquid state, in which it assumes the shape of its container, to a plastic state, in which the pliable soil mass retains its formed shape. In the transition plastic water content is called the *liquid limit, LL*. As the soil dries further, but still is saturated, it passes from the plastic state to a semi-solid state, in which it is no longer pliable. This transition water content is called the *plastic limit, PL*. Then, as further drying occurs, the soil finally reaches a state in which it becomes partially saturated and brittle. In this transition

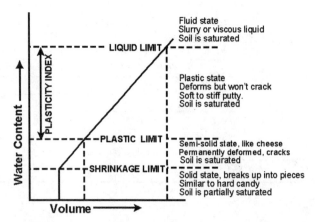

FIGURE 2.2 Atterberg limits—comparison of water content and volume change.

state, the water content has reached the *shrinkage limit, SL*. The difference in water content between the liquid limit and the plastic limit is called the *plasticity index, PI;* that is:

$$PI = LL - PL \tag{2.3}$$

Because the transition from one state to another is distinct rather than gradual, the definitions of the terms are based on water contents that are determined by the following highly standardized tests (ASTM D 4318). The Atterberg limits tests are made only on that portion of the total soil sample that is finer than the No. 40 (0.42 mm) screen.

Liquid Limit Test. A patty of the wet soil sample is formed in a small, shallow, metal cup and a groove is cut in the pat. The cup is impacted in a standard manner until the groove closes for a distance of 13 mm (½ inch). This procedure is repeated as more water is added, and the water content, after 25 such impacts is the *liquid limit*.

Plastic Limit Test. A portion of the soil sample is rolled into a thread in trials at various water contents. The water content at which the thread just begins to crumble when it reaches 3.2 mm (⅛ inch) in diameter is the *plastic limit*.

Shrinkage Limit Test. A portion of the soil sample that is at a water content above the liquid limit is placed in a container of known volume. The sample is dried slowly in an oven at a constant temperature of 105°C until it reaches constant weight (fully dried). The volume of the remaining sample is measured, and the water content at which shrinkage stopped is calculated as the *shrinkage limit*.

The relationship between water content and volume of the cohesive soil mass is shown in Figure 2.2. For a given mass of soil plus water, drying simply reduces the amount of water in the mass. Therefore, while the soil is fully saturated, the change in volume of its mass is equal to the change in volume of the water that has been removed. At the shrinkage limit, the volume change stops, and the sample becomes partially saturated as air moves into the voids. The volume change because of a plasticity index water content change also is indicated in Figure 2.2.

Liquidity Index. The relationship between consistency (stiffness) of a remolded clayey soil and the Atterberg limits is defined as the *liquidity index, LI*:

$$LI = \frac{w - PL}{LL - PL} = \frac{w - PL}{PI} \tag{2.4}$$

When the liquidity index equals 1, the water content of the soil is at the liquid limit, and the soil is very soft. When the liquidity index equals 0, the water content of the soil is at the plastic limit, and the soil is stiff or hard. This applies only to saturated soil in the remolded state.

Activity Index. The various clay mineral types described above have different volume changes from the liquid limit to the plastic limit for equal amounts of clay sizes in the fraction finer than the No. 40 (0.42 mm) sieve. This behavior has been defined [4] as the *activity* of the clay. When the clay content, the percentage of the fraction finer than 0.42 mm

(No. 40 screen) that is finer than 0.002 mm (2 micron) obtained from the hydrometer test part of the grain size distribution graph of Figure 2.1, is greater than 40 percent, the activity index AI is defined as:

$$AI = \frac{PI}{\% < 0.002 \ mm} \qquad (2.5)$$

When the clay content is less than 40 percent the activity index becomes [5]:

$$AI = \frac{PI}{\% < 0.002mm - 5} \qquad (2.6)$$

The activity indices for various clays are (a) kaolinite, 0.3 to 0.5, (b) illite, 0.5 to 1.3, (c) Ca-montmorillonite, about 1.5, and (d) Na-montmorillonite, 3 to 7. Intermediate values indicate a mixture of clay mineral types. A high activity index is associated with those clay minerals that can absorb large amounts of water within their crystal lattice and is a function of the chemistry of the clay particles. Clays with an activity index below 0.75 are defined as *inactive* clays. Those with an activity index between 0.75 and 1.25 are *normal activity* clays, and those with an activity index greater than 1.25 are defined as *active* clays.

Atterberg limits and clay content. Generally, the liquid limit for a given soil is directly proportional to the clay content because silt and sand grains between 0.42 mm (No. 40) and 0.002 mm act as inert fillers. The plastic limit, however, is directly proportional to the clay content only above between 40 percent to 45 percent clay. The plastic limit is inversely proportional to the clay content below that percentage. Since the plasticity index depends on the difference between the liquid and the plastic limits, the relationship with the activity index ideally should apply only to clay-rich soils.

In the case of a granular soil—one having a high percentage of particles retained on the No. 40 screen (since the Atterberg limits are determined only on the fraction passing the No. 40 screen)—the percentage of clay is determined on the basis of the whole sample. Therefore, the calculation of the activity index using the *whole sample percent clay* is inconsistent. To resolve this, only the clay determined as a fraction of the portion of the whole sample smaller than the No. 40 screen should be considered. For example, a clayey, gravelly sand with $PI = 15$ (20 percent finer than No. 40 and 5 percent finer than 0.002 mm) would have an adjusted percentage of clay finer than 0.0002 mm equal to 5/20 or 25 percent of the smaller-than No. 40 fraction. The adjusted activity of the clayey, gravelly sand would then become $AI = 15/25 = 0.6$, indicating a true low activity.

SOIL CLASSIFICATION SYSTEMS

Soils are classified according to criteria that are based on specific identification tests and observations. A soil classification system, as used in geotechnical engineering practice, is a grouping or rating that provides an indication of expected or probable behavior in a specific engineering application. Soil classification is *interpretive* information, whereas soil identification is *factual* information.

Three soil classification systems are found in the geotechnical engineering and related literature: (a) the Unified Soil Classification System (USCS), (b) the American Association of State Highway and Transportation Officials (AASHTO) Highway Soil Classification System, and (c) the United States Department of Agriculture (USDA) Soil Taxonomy. The USCS and AASHTO systems were developed to serve special engineering- or construction-related purposes. The USDA Soil Taxonomy system was developed for use in agricultural soil science.

Unified Soil Classification System

The USCS was derived from the Airfield Classification System (ACS) developed in 1942 by Dr. Arthur Casagrande to facilitate wartime military airfield construction by the U.S. Army Corps of Engineers. Its purpose was to permit the categorization of soils according to their desirability for use in airfield base construction. In the early 1950s, several U.S. government agencies involved in soils engineering agreed on a unified soil classification system based on the ACS. In 1953, the USAE Waterways Experiment Station (WES) published the USCS as a technical report, which was updated in 1960 [6]. The USCS later was adopted by the American Society For Testing and Materials (ASTM) as Standard D 2487[3].

The USCS, shown in Tables 2.2 and 2.3, distinguishes two major categories of soil, based on grain size:

1. Coarse-grained soils (Table 2.2). Mineral soil particles more than 50 percent by weight coarser than (particles retained on) the U.S. No. 200 screen (0.074 mm).
2. Fine-grained soils (Table 2.3). Mineral soil particles with 50 percent or more by weight finer than (those particles passing) the U.S. No. 200 screen (0.074 mm).

All soil types are divided into soil groups, each of which must meet unique criteria and is described by a group name and a two-letter group symbol.

Coarse-Grained Soils. Coarse-grained soils are subdivided into two groups, based on grain size:

1. Gravel. More than 50 percent by weight of the coarse fraction (retained on No. 200) is coarser than the U.S. No. 4 screen (4.75 mm).
2. Sand. At least 50 percent by weight of the coarse fraction (retained on No. 200) is finer than the U.S. No. 4 screen (4.75 mm).

The coefficients of uniformity C_u and curvature C_z are calculated from a grain-size distribution graph using Equations 2.1 and 2.2.

Fine-Grained Soils. Fine-grained soils are grouped according to plasticity. The liquid limit and plasticity indexes are determined by the ASTM D 4318 standard procedure [3].

The USCS uses a soil plasticity graph, Figure 2.3, to distinguish two fine-grained soil types:

1. High plasticity soils: LL is greater than or equal to 50 percent.
2. Low plasticity soils: LL is less than 50 percent.

TABLE 2.2 Group Symbols for Gravelly and Sandy Soils. Unified Soil Classification System [3]

Group Symbol	Group name*	Criteria	
		Percent passing No. 200 sieve	Other factors [†]
GRAVELLY SOILS—More than 50% retained on No. 200 sieve. More than 50% of coarse fraction retained on No. 4 sieve.			
GW	Well-graded gravel	< 5	C_u greater than or equal to 4; C_c between 1 and 3.
GP	Poorly graded gravel	< 5	Not meeting both criteria for GW.
GM	Silty gravel	> 12	Atterberg limits plot below "A" line or plasticity index less than 4.
GC	Clayey gravel	> 12	Atterberg limits plot on or above "A" line and plasticity index greater than 7.
GC-GM	Silty, clayey gravel	> 12	Atterberg limits plots on or above "A" line and plasticity index is between 4 and 7.
GW-GM	Well-graded gravel with silt	5–12	Meets criteria for GW and GM.
GW-GC	Well-graded gravel with clay	5–12	Meets criteria for GW and GC.
GP-GM	Poorly graded gravel with silt	5–12	Meets criteria for GP and GM.
GP-GC	Poorly graded gravel with clay	5–12	Meets criteria for GP and GC.
SANDY SOILS--More than 50% retained on No. 200 sieve. 50% or more of coarse fraction passes No. 4 sieve.			
SW	Well-graded sand	< 5	C_u greater than or equal to 6; C_c between 1 and 3.
SP	Poorly graded sand	< 5	Not meeting both criteria for SW.
SM	Silty sand	> 12	Atterberg limits plot below "A" line or plasticity index less than 4.
SC	Clayey sand	> 12	Atterberg limits plot on or above "A" line and plasticity index greater than 7.
SC-SM	Silty, clayey sand	> 12	Atterberg limits plots on or above "A" line and plasticity index is between 4 and 7.
SW-SM	Well-graded sand with silt	5–12	Meets criteria for SW and SM.
SW-SC	Well-graded sand with clay	5–12	Meets criteria for SW and SC.
SP-SM	Poorly graded sand with silt	5–12	Meets criteria for SP and SM.
SP-SC	Poorly graded sand with clay	5–12	Meets criteria for SP and SC.

*If sample contains 15 % or more gravel, add "with gravel" to group name.
If fines are organic, add "with organic fines" to group name.

[†] C_u = Coefficient of Uniformity = D_{60} / D_{10}
C_c = Coefficient of Curvature = $(D_{30})^2 / (D_{10} \times D_{60})$

TABLE 2.3 Group Symbols for Silty and Clayey Soils. Unified Soil Classification System [3]

Group symbol	Group name*	Criteria	
		Liquid Limit	Other factors
CL	Lean clay	Less than 50	Plasticity index plots on or above "A" line and PI is greater than 7.
CL-ML	Silty clay		Plasticity index plots on or above "A" line and PI is between 4 and 7.
ML	Silt		Plasticity index plots below "A" line or PI is less than 4.
OL [†]	Organic clay		Plasticity index plots on or above "A" line and PI is equal to or greater than 4.
	Organic silt		Plasticity index plots below "A" line or PI is less than 4.
CH	Fat clay	Equal to or greater than 50	Plasticity index plots on or above "A" line.
MH	Elastic silt		Plasticity index plots below "A" line.
OH [†]	Organic clay		Plasticity index plots on or above "A" line.
	Organic silt		Plasticity index plots below "A" line.
Pt	Peat	Highly organic soils.	Primarily organic matter, dark in color, and organic odor.

*If sample contains 15 to 29 % plus No. 200, add "with sand" or "with gravel" to group name, whichever is predominant.
If sample contains 30 % or more plus No. 200, add "sandy" or "gravelly" to group name, whichever is predominant.

[†] Organic soils have (Liquid limit - oven dried) / (Liquid limit - not dried) < 0.75

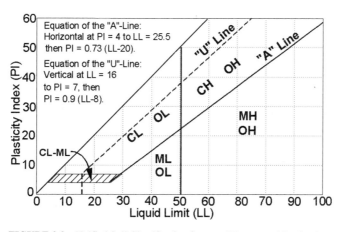

FIGURE 2.3 Unified Soil Classification System. (*Courtesy of Portland cement association*).

Organic soils can be distinguished from *inorganic* soils by means of the liquid limit test. The standard liquid limit test is made on a soil sample that has not been dried previously. The test is repeated on the sample after it has been oven-dried. For organic soils, the ratio of the liquid limit *after* oven drying to the liquid limit *before* oven drying is less than 0.75.

AASHTO Highway Soil Classification System

Transportation departments in the United States use the AASHTO soil classification system [7] to classify soils for use as highway and airfield embankments, subgrades, sub-bases, and bases. Like the USCS, AASHTO uses both gradation and Atterberg limits for its classification criteria, as shown in Table 2.4.

To use the AASHTO classification system, proceed as follows:

1. Make a mechanical sieve analysis test (see Figure. 2.1) and Atterberg limits tests.
2. Using the results of the sieve analysis, consult Table 2.4, proceeding from left to right (i.e., from type A-1 to A-2). The first group that fits the test data is the correct classification.
3. Calculate the group index GI value from the equation

$$GI = (F - 35) \times [0.2 + 0.005 \times (LL - 40)] + 0.01 \times (F - 15) \times (PI - 10) \qquad (2.7)$$

where F is the percentage passing through the No. 200 sieve (0.074 mm), expressed as a whole number. This is based only on the fraction of the material passing the 75 mm (3-inch) screen. When the calculated group index is negative, the GI is reported as zero. When calculating the group index for subgroups A-2-6 and A-2-7, only the PI portion of Equation 2.7 is used.

4. The results are presented with the group index values in parentheses after the group symbol, for example, A-2-6(3), A-4(5), or A-7-5(17).

USDA Soil Taxonomy

The USDA Soil Conservation Service (renamed the Natural Resources Conservation Service in 1994) adopted in 1975 a revised soil classification system titled *Soil Taxonomy: A Basic System of Soil Classification for Making and Interpreting Soil Surveys* [8]. The basic agronomic soil mapping unit is the *soil series*, whose members have the same genesis and weathering profile. This implies that soils that have the same kind of parent material, climate, and native vegetation also have the same number of horizons of similar depth, have essentially the same slope and landscape position, and are of about the same geologic age. Series having similar but not identical characteristics are grouped into *families*. Similar families are grouped into *subgroups*, then into *great groups*, and then into *suborders*. The highest category of the Soil Taxonomy system is the *order*, of which 10 have been defined.

Identifying Characteristics of Soils. The identifying characteristics of the soil of each horizon layer consist of (a) color, (b) texture, (c) structure, and (d) consistency. These characteristics roughly correspond to the soil's engineering properties. The texture is a measure

TABLE 2.4 AASHTO Classification of Soils [7]

Group classification	A-1		A-3	A-2				A-4	A-5
	A-1-a	A-1-b		A-2-4	A-2-5	A-2-6	A-2-7		
Sieve analysis: percent passing:									
2.00 mm (No. 10)	50 max.	----	----					----	----
0.425 mm (No. 40)	30 max.	50 max.	51 min.	----	----	----	----	----	----
0.074 mm (No. 200)	15 max.	25 max.	10 max.	35 max.	35 max.	35 max.	35 max.	36 min.	36 min.
Characteristics of fraction passing 0.425 mm (No. 40):									
Liquid Limit	----		----	40 max.	41 min.	40 max.	41 min.	40 max.	41 min.
Plasticity Index	6 max.		N.P.	10 max.	10 max.	11 min.	11 min.	10 min.	10 max.
Usual types of significant constituent materials.	Stone fragments, gravel, and sand.		Fine sand	Silty or clayey gravel and sand.				Silty soils.	
General rating as subgrade.	Excellent to good								

* Plasticity index of A-7-5 subgroup is equal to, or less than, LL minus 30. Plasticity index of A-7-6 subgroup is greater than LL minus 30.

of material grain properties (grain size). The structure corresponds to the mass properties (density), and the consistency corresponds to the physical behavior properties (strength).

Soil Texture. The basic soil textural class names used by the USDA [8] are defined in terms of the relative amounts of sand, silt, and clay (based only on the percentage passing through the No. 4 screen) determined by laboratory grain size analysis, as shown in Table 2.5. Gravel and coarser sizes are not included in the definitions.

ROCK IDENTIFICATION AND CLASSIFICATION

Rock classification for engineering purposes consists of two basic assessments: (a) *intact character*, that is, the characteristics of intact rock fragments or cores, and (b) *in situ character*, that is, the engineering features of in-place rock masses.

Intact Character

A typical description of a rock sample, fragment, or core, is based on visual and manual observations. The description includes the following [7] [9]:

1. Weathering. The sample is described as fresh, slightly weathered, and so on, in accordance with Table 2.6.
2. Color and texture. Color is described in accordance with basic colors on any standard color description system. Size, shape, and arrangement of constitutive elements are described in accordance with Table 2.7.
3. Hardness. The sample is described as very soft, soft, hard, and so on, in accordance with the classes contained in Table 2.8.
4. Geological name. The rock is identified by its geologic name and local name, if it has one. Subordinate constituents in the rock sample, such as seams or bands of other mineral types, are included. The identification and geologic naming of rocks is an extensive and detailed subject, beyond the scope of this book. The reader is referred to standard textbooks on geology or rock identification, or to a qualified geologist. An example of a typical description based on a visual-manual observation is "slightly weathered brown fine-grained hard silt stone."

In Situ Character

The engineering characteristics of an *in situ* rock mass generally are concerned with its structural elements, especially its discontinuities. Fractures in exposed rock surfaces are described in terms of frequency, attitude, spacing, roughness, bonding quality, and general continuity. The typical description of an *in situ* rock mass can include, in addition to the visual-manual descriptions given above, the following:

1. Rock Quality Designation (RQD). The RQD was proposed by Deere [10] as an indicator of the frequency of discontinuities (fractures) in a rock mass. In general, the

TABLE 2.5 Textural Classification of Soils. [8]
U. S. Department of Agriculture, Natural Resource Conservation Service

TEXTURE CLASS	COMPOSITION (%) SAND	SILT	CLAY	PARTICLE SIZES AND VISUAL APPEARANCE
SANDY SOILS—Coarse-textured soils.				
Sand	85–100	0–15	0–10	Loose and single-grained. Individual grain sizes can be detected. Free-flows when dry.
Loamy Sand	70–90	0–30	0–15	Granular soil with sufficient silt and clay to make it somewhat plastic. Sand character predominates.
LOAMY SOILS—Moderately coarse-textured soils.				
Sandy Loam	43–85	0–50	0–20	Granular soil with sufficient silt and clay to make it somewhat coherent. Sand character predominates.
LOAMY SOILS—Medium-textured soils.				
Loam	23-52	28-50	7-27	Uniform mix of sand, silt, and clay. Uniform sand gradation from coarse to fine. Slightly gritty feel, but smooth and plastic.
Silt Loam	0-50	50-88	0-27	Small amount of fine sand and/or clay. Cloddy when dry. Friable, readily broken and pulverized.
Silt	0-20	80-100	0-12	Very little fine sand and/or clay. May be cloddy when dry. Friable, readily broken and pulverized.
LOAMY SOILS--Moderately fine-textured soils.				
Clay Loam	20-45	15-53	27-40	Fine-textured. Makes hard lumps when dry. Resembles clay when in dry condition.
Sandy Clay Loam	45-80	0-28	20-35	Granular soil with sufficient clay to make it somewhat plastic and coherent. Sand character is somewhat masked.
Silty Clay Loam	0-20	40-73	27-40	Very little fine sand. May be cloddy when dry. Somewhat friable. Broken and pulverized with some effort.
CLAYEY SOILS--Fine-textured soils.				
Sandy Clay	45-65	0-20	35-55	Clayey soil with sufficient sand to make it somewhat friable.
Silty Clay	0-20	40-60	40-60	Clay and silt mixture. Sufficient clay to make it somewhat smooth and plastic.
Clay	0-45	0-40	40-100	Clayey soil. Makes hard lumps when dry. Not friable; difficult to crumble into powder when dry.

RQD is defined as the total length of NX-size (54 mm, 2-⅛ in. ID) or larger core samples that are recovered from a borehole and are greater than or equal to 10 cm (4 in.) in length, divided by the total length of the core run.

TABLE 2.6 Rock Weathering Description.

Grade	Diagnostic Features
Fresh	No visible sign of decomposition or discoloration. Rock rings under hammer if crystalline.
Slightly Weathered	Slight discoloration along discontinuities–otherwise same as "Fresh."
Moderately Weathered	Significant portions of rock show discoloration and weathering effects. Surface pitted. Strength somewhat less than fresh rock but cores cannot be easily broken by hand or easily scraped by a knife.
Highly (Severely) Weathered	Entire mass discolored. Most minerals somewhat decomposed. Specimen can be broken by hand with effort or shaved with a knife. Rock gives "clunk" sound when struck with hammer. Core stones present in rock mass. Texture becoming indistinct.
Completely Weathered (Decomposed)	Rock reduced to *soil*. Rock fabric and structure are not discernable. Specimens are easily crumbled or penetrated.

TABLE 2.7 Rock Textural Description.

Grain Size	Diagnostic Features
Very Coarse Grained	Particles greater than 6 mm (0.25 in.)
Coarse Grained	Mineral grains between 2.5 mm (0.1 in.) and 6 mm (0.25 in.)
Medium Grained	Mineral grains barely detectable with naked eye; to 2.5 mm (0.1 in.)
Fine Grained	Constituent mineral grains barely detectable with naked eye
Very Fine Grained	Constituent mineral grains too small to be seen with naked eye
Glassy	Smooth, shiny surface; no grains detectable even with 10X hand lens

TABLE 2.8 Rock Hardness Description.

Hardness	Field Test	Approximate range of uniaxial compressive strength	
		MPa	tons/sq. ft.
Extremely Hard	Many blows with geologic hammer required to break intact specimen. Cannot be scratched with knife.	>200	> 2000
Very Hard	Hand-held specimen breaks with hammer end of pick under a few blows (more than one blow). Cannot be scratched with knife.	200–100	2000–1000
Hard	Cannot be scratched, scraped, or peeled with knife; hand-held specimen can be broken with single moderate blow with pick.	100–50	1000–500
Soft	Material crumbles under moderate blow with sharp end of pick. Can be scratched by fingernail and gouged or grooved readily by knife or pick.	50–25	500–250
Very Soft	Material crumbles under moderate blow with sharp end of pick and can be peeled with knife. Pieces 25 mm (1 in.) or more in thickness can be broken by finger pressure.	25–1	250-10

2. Seismic velocity and/or velocity index. The seismic velocity of a rock can be determined in the field and laboratory by measuring the velocity of a compressional wave created by an energy source. Field and laboratory velocities are determined by geophysical techniques using seismic sources and pickups. The velocity index is defined as the square of the ratio of the *field* compressional wave velocity to the *laboratory* compressional wave velocity.

3. Unconfined compressive strength and point load strength. The strength of intact rock specimens can be determined by either the unconfined compression test or the point load test. The unconfined (uniaxial) compression test is made on an intact, cylindrical section of core that has a height two or more times the diameter and has plane ends. The point load test is an indirect method of determining unconfined compressive strength. The test can be made either in the field or in the laboratory on core pieces too irregular for the unconfined compression test. A general but approximate relationship between unconfined compressive strength UC and point load strength PLS is as follows:

$$UC \approx 25 \times PLS \tag{2.8}$$

Approximate values of the ratio between field-determined and laboratory-determined tests of the modulus of elasticity and/or unconfined compressive strength are given in Table 2.9.

These values should be used only as general indicators and should be checked for each specific project. Qualitative relationships between rock mass quality, RQD, the velocity index, and the ratio between field and laboratory determinations of the modulus of elasticity and of unconfined compressive strength also are given in Table 2.9.

WEIGHT-VOLUME (PHASE) RELATIONSHIPS

Each mass of soil and rock contains some amount of each of three components: solids, liquid, and gas. The interrelations of the weights (or masses) and volumes of the different phases help define the condition or physical makeup of the material. The liquid is ordinarily water and the gas is ordinarily air, although the water can contain dissolved salts, and the gas can be methane in some soils containing decaying matter.

TABLE 2.9 Engineering Description for Rock Quality.

Rock mass quality	RQD %	Velocity Index V_F / V_L	Estimated field/lab. Unconfined Compression ratio, UC_F / UC_L
Excellent	90–100	0.80–1.00	0.7–1.0
Good	75–90	0.60–0.80	0.3–0.7
Fair	50–75	0.40–0.60	0.25
Poor	25–50	0.20–0.40	0.20
Very Poor	0–25	0–0.20	0.15

For definition and calculation purposes, it is assumed that the solids, liquid, and gas are separated as shown in Figure 2.4; that is, all of the solid soil or rock particles are assumed to occupy a space that is free of liquids or gases, and the liquid is assumed to be free of any gases. With this stipulated, various terms and relationships can be determined from Figure. 2.5. In the English system (foot, pound, second), *force* is a fundamental unit and the weight of rock or soil is used in the calculations. In the International System of Units (SI/metric), *mass* is a fundamental unit, and mass is used.

EXAMPLE PROBLEMS

Grain Size Distribution

2.1 Question:
Refer to the grain size curve for well-graded fine sand in Figure 2.1. Does this sample meet all the criteria for a well-graded sand?

Answer:
The 60 percent size, D_{60}, is 0.4 mm; the 30 percent size, D_{30}, is 0.13 mm; and the 10 percent size, D_{10}, is 0.04 mm. The coefficient of uniformity (Equation 2.1) is $C_u = 0.4 / 0.04 = 10$, and the coefficient of curvature (Equation 2.2) is $C_z = 0.13^2 / (0.4 \times 0.04) = 1.06$. Both values meet the criteria for a well-graded sand.

2.2 Question:
Refer to the grain-size curve for uniform fine sand in Figure. 2.1. Does this sample meet all the criteria for well-graded sand?

Answer:
The 60 percent size, D_{60}, is 0.19 mm; the 30 percent size, D_{30}, is 0.14 mm; and the 10 percent size, D_{10}, is 0.095 mm. The coefficient of uniformity is $C_u = 0.19 / 0.095 = 2$, and the coefficient of curvature is $C_z = 0.14^2 / (0.19 \times 0.095) = 1.09$. The value of C_u for this sand fails the uniformity criterion for well-graded sand.

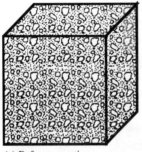
(a) Before separation

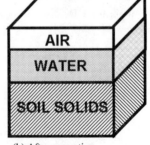
(b) After separation

FIGURE 2.4 Unit cubes of soil before and after imaginary separation of phases.

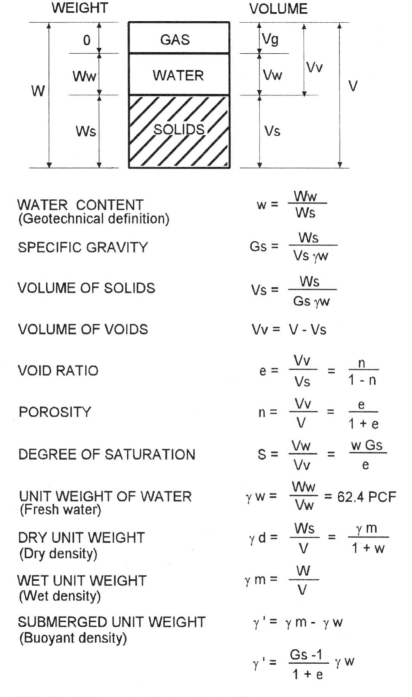

FIGURE 2.5 Definitions of weight-volume (phase) relationship terms.

2.3 Question:
Refer to the grain-size curve for gap-graded sand in Figure. 2.1. Does this sample meet all the criteria for well-graded sand?

Answer:
The 60 percent size, D_{60}, is 6 mm; the 30 percent size, D_{30}, is 0.34 mm; and the 10 percent size, D_{10}, is 0.095 mm. The coefficient of uniformity is $C_u = 6 / 0.095 = 63$, and the coefficient of curvature is $C_z = 0.34^2 / (6 \times 0.095) = 0.2$. The value of C_z for this material fails the curvature criterion for well-graded sand.

Atterberg Limits

2.4 Question:
A saturated, remolded clay soil has a water content of 32 percent, a liquid limit water content of 35 percent, a plastic limit water content of 20 percent, and an amount of clay sizes (finer than 0.002 mm) of 17 percent. (a) What are the liquidity index and activity index of this soil? (b) Is the sample likely to be soft or stiff? (c) What is the activity index for this soil and which clay minerals does it possess?

Answer:

a. The plasticity index (Equation 2.3) of this clayey soil is PI = 35 − 20 = 15. The liquidity index (Equation 2.4) is LI = (32 − 20) / (35 − 20) = 12 / 15 = 0.8.

b. Because the liquidity index is near one, the sample will be soft.

c. The activity index (Equation 2.5) is AI = 15 / (17 − 5) = 1.25. This indicates that the dominant clay mineral is an illite, or that the clay is a mixture of several clay minerals including kaolinite, illite, and montmorillonite.

2.5 Question:
If the saturated, remolded clay soil of Problem 2.4 is dried to a water content of 22 percent, (a) what is its liquidity index? (b) Is the sample likely to be soft or stiff?

Answer:

a. The liquidity index is LI = (22 − 20) / (35 − 20) = 2 /15 = 0.13.

b. Because the liquidity index is near zero, the sample will be stiff to hard.

Soil Classification

2.6 Question:
Plot the grain size data in Table 2.10 on the graph of Figure 2.6 and the Atterberg limits data on the graph of Figure 2.6. Then fill in and answer the questions posed in Table 2.11.

Answer:
Place your solutions to the several questions about the data of Problem 2.6 in the appropriate spaces of the worksheet, Table 2.11. Verify all solutions.

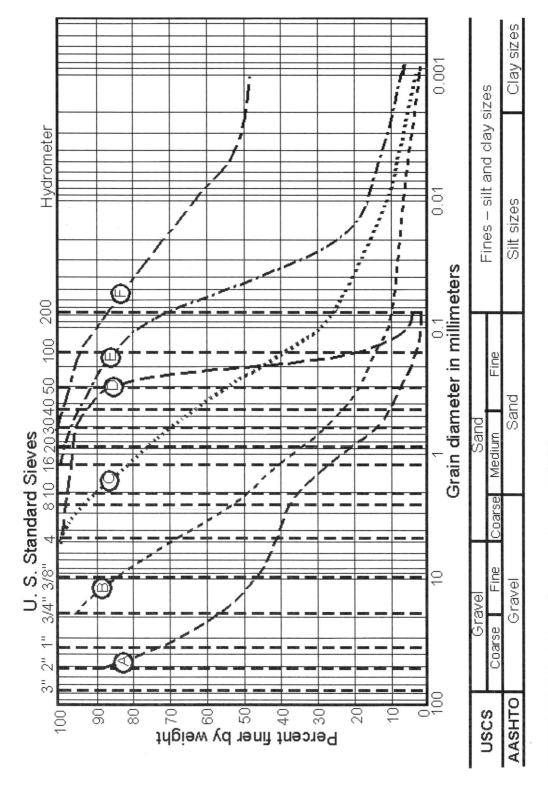

FIGURE 2.6 Typical grain size distribution curves (accompanies E.P.2.6).

TABLE 2.10 Data for Example Problem 2.6.

Sample	A	B	C	D	E	F
	Percent passing (finer than) sieve size shown.					
2 in.	88	-----	-----	-----	-----	-----
1 in.	73	-----	-----	-----	-----	-----
¾ in.	58	96	-----	-----	-----	-----
⅜ in.	47	84	-----	-----	-----	-----
No. 4	41	68	100	100	-----	-----
No. 8	38	52	94	98	-----	-----
No. 16	28	41	82	97	99	-----
No. 30	15	28	70	95	97	100
No. 50	8	19	55	87	91	97
No. 100	5	15	39	20	86	95
No. 200	2	9	26	4	72	88
0.010 mm*	-----	7	11	-----	25	63
0.005 mm*	-----	6	8	-----	14	52
0.001 mm*	-----	3	4	-----	7	48
Liquid Limit	NP†	35	25	NP	46	72
Plastic Limit	NP†	21	19	NP	24	47

* From hydrometer test
† Non-Plastic

Rock Description and Classification

2.7. Question:
A length of core is retrieved from a drill hole with an NX-size core barrel. A piece of rock core is examined and the following observations are made: (a) it is gray in color; (b) it is limestone (fizzes with weak hydrochloric acid); (c) more than half of the rock core shows some discoloration, and the surface is slightly pitted; (d) individual grains cannot be seen but the surface is not glassy; and (e) the core can barely be scratched with a knife but can be broken with a sharp blow of a hammer. The entire length of core-run is examined and all intact pieces 10 cm (4 in.) or more in length are measured along their centerlines. The total of all the measured intact pieces is 116 cm (45.8 in.). The total length of the recovered core is 142 cm (56 in.). (a) What is the visual-manual description of this rock core? (b) What is the RQD for this length of core?

Answer:

a. Using Tables 2.6, 2.7, and 2.8, one can describe the sample as "moderately weathered, gray, very fine-grained, hard limestone."

TABLE 2.11 Solutions for Example Problem 2.6.

Sample	A	B	C	D	E	F
60 percent size, D_{60}, mm	21	4.4	0.48	0.22	0.056	-----
30 percent size, D_{30}, mm	1.4	0.7	0.10	0.18	0.024	-----
10 percent size, D_{10}, mm	0.4	0.075	0.008	0.13	0.002	-----
Coefficient of uniformity, C_u (Eq. 2.1)	52	59	60	1.7	28	-----
Coefficient of curvature, C_z (Eq. 2.2)	0.23	1.5	2.6	1.1	5.1	-----
Is the sample well-graded?	Yes	Yes	Yes	No	Yes	-----
Is the sample uniform?	No	No	No	Yes	No	-----
Is the sample gap-graded?	Yes	No	No	No	-----	-----
What is the sample's plasticity index (Eq. 2.3)?	NP	14	6	NP	22	25
What is the percent finer than No. 40?	10	23	64	93	95	98
What is the percent clay (<0.002 mm) as a fraction of the whole sample?	-----	4	6	-----	10	49
What is the percent clay (<0.002 mm) as a fraction of the percent < No. 40?	-----	17.4	9.4	-----	10.5	50
What is the "adjusted" Activity Index (Eq. 2.5 for clay >40%, Eq. 2.6 for clay <40%)? *	-----	1.9	0.9	-----	1.8	0.47
What is the percent retained on No. 200?	98	91	74	96	28	12
Of this amount, what percent passes No. 4?	40	65	100	100	NA †	NA †
What is the USCS Classification?	GP	SW-SC	CL-ML	SP	CL	MH
What is the AASHTO Classification?	A-1-a	A-2-6	A-2-6	A-3	A-6	A-7-5
What is the AASHTO Group Index?	0	0	0	0	12	29
Of the percentage passing the No. 4,						
What percent is sand?	95	87	74	96	28	12
What percent is silt?		7	20		62	39
What percent is clay?		95				
What is the USDA Agricultural Classification for this sample?	Sand and Gravel	Sand	Loamy Sand	Sand	Silty Clay	Loam

* Percent clay (< 0.002 mm) calculated as percent of the < No. 40 (0.42 mm) fraction.

† Not Applicable

b. The RQD is defined as the sum total length of all intact pieces 10 cm (4 in.) or more in length divided by the total length of the core-run, expressed as a percentage. Therefore RQD = 100 (116 / 142) = 82 %. Data from Table 2.9 indicates that this core-run rates as a rock mass of *good* quality.

Weight-Volume (phase) Relationships

In each of the following weight-volume example problems, enter all known data in the accompanying block diagram, make the necessary calculations, and enter the resulting weight and volume data on the diagram. Assume a cube volume of 1.0 in all problems. Show all calculations.

2.8 Question:

A sample of sandy soil is removed from the ground surface. The wet soil (solids plus water) weighs 15.0 kg. (33.04 lb.). The volume of the hole is measured to be 7.35 dm^3 = 7.35 liters (0.26 cu. ft.). A sample of the soil is dried at 105°C to constant temperature. The weight of the sample before drying is 450 g; after drying it weighs 385 g. The specific gravity of the soil solids is determined to be Gs = 2.68. Calculate (a) the wet density, (b) the percentage of water content, (c) the dry density, (d) the degree of saturation, (e) the saturating water content, (f) the porosity, and (g) the void ratio. Show all values in appropriate locations on Figure 2.7.

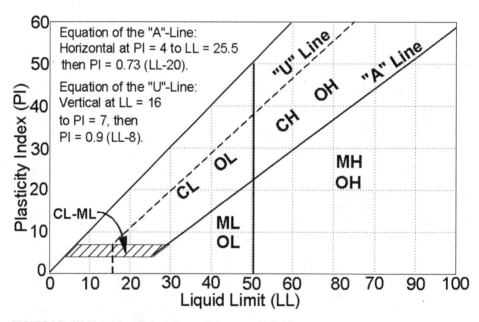

FIGURE 2.7   USCS Atterberg limits A-line graph (accompanies E 2.6).

Answer:

a. The wet unit weight:
$W/V = 15.0 / 7.35 = 2041$ g/l
$W/V = 33.04 / 0.26 = 127.1$ ft³

b. The water content:
$w\% = 100 \times (450 - 385) / 385 = 16.9\%$

c. The dry density:
$d = Ws / V = 2041 / (1 + 0.169) = 1746$ g/l
$d = Ws / V = 127.1 / (1 + 0.169) = 108.7$ ft³

d. The volume of solids: $Vs = Ws / Gs = 1746 / (2.68 \times 1000) = 0.651$ l
$Vs = Ws / Gs = 108.7 / (2.68 \times 62.4) = 0.651$ ft³
The volume of voids: $Vv = 1 - Vs = 1 - 0.651 = 0.349$ liter $= 0.349$ ft³
The volume of water: $Vw = Ww / Gs = (2041 - 1746) / 1000 = 0.295$ l
$Vw = (127.1 - 108.7) / 62.4 = 0.295$ ft³
The degree of saturation: $S\% = Vw / Vv = 100 \times 0.295 / 0.349 = 84.5\%$

e. The saturating water content: $w(sat)\% = Ww / Ws$ when $Vw = Vv$
$Vw = Vv = 0.349$ and $Ww = 349$ g/l
$w\% = 100 \times (349 \times 1.0) / 1746 = 20.0\%$
$Vw = Vv = 0.349 \times 62.4 = 21.8$ lb/ft³
$w\% = 100 \times 21.8 / 108.7 = 20.0\%$

f. The porosity: $n = Vv / V = 0.349 / 1.0 = 0.349$

g. The void ratio: $e = Vv / Vs = 0.349 / 0.651 = 0.536$

2.9 Question:
A laboratory test is made by adding sufficient water to a sample of remolded clayey soil to bring its consistency above the liquid limit water content. The wet soil is placed in a stainless steel container with an inside volume of 100 cc. The sample of the soil is dried at 105°C. to a constant temperature. The weight of the sample before drying (solids plus water) is 185 g (0.407 lb). After drying, it weighs 135 g (0.297 lb) and its measured volume is 67 cc. The specific gravity of the soil solids was determined in a separate test to be Gs = 2.70. (a) What is the shrinkage limit for this soil? (b) If the liquid limit is determined to be 36 percent and the plastic limit is determined to be 17 percent, how much volume change, in percent, would one volume unit (liter, cubic foot) change if the water content changes from the liquid limit to the plastic limit? Show all values in appropriate locations on Figure 2.8. Assume a block of soil 1 cm wide and 1 cm deep and 100 cm high, for a volume of 100 cc.

(*Hint*: Figure 2.2 shows that the volume change from above the liquid limit water content down to the shrinkage limit is directly, linearly related to the change in the amount of water in the sample from oven drying. Although the water content continues to lower to zero below the shrinkage limit, there is no further volume change and air enters the void space previously occupied by the water.)

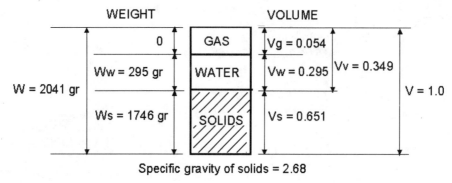

FIGURE 2.8 Block diagram (accompanies E 2.8).

Answer:
a. The volume of solids for 135 g of soil is the same before and after drying:
$Vs = Ws / Gs = 135 / 2.70 = 50$ cc
b. The total weight of the sample before drying is 185 g and after drying is 135 g. Therefore, the weight of water lost from drying is $185 - 135 = 50$ g, and the water content before drying is $Ww / Ws = 100 \times 50 / 135 = 37\%$. These values are shown in Figure 2.9(a). After drying, the total volume is 67 cc and the volume of solids is 50 cc. Therefore, the volume of voids after drying is $Vv = 67 - 50 = 17$ cc, which is now filled with air. At the shrinkage limit, the volume of voids still is filled with water. Therefore, the volume of water Vw equals the volume of voids Vv which equals 17 cc. At that point, the water content (the shrinkage limit) is $SL = 100 \times 17 / 135 = 12.6\%$. These values are shown in Figure. 2.9(b).

2.10. Question:
The soil of Problem 2.9 has a liquid limit water content of 36 percent and a plastic limit water content of 18 percent. What is the volume change, in percent, for this saturated, remolded soil as the water content changes from the liquid limit to the plastic limit?

Answer:
As shown in Figure 2.9, at the liquid limit the weight of water is $Ww = w \cdot Ws = 0.36 \times 135 = 48.6$ g, which occupies a volume of 48.6 cc (Vw), and has a total volume of 98.6 cc ($Vs + Vw = 50 + 48.6$). At the plastic limit the weight of water is $Ww = w \times Ws = 0.18 \times 135 = 24.3$ g, which occupies a volume of 24.3 cc (Vw), and a total volume of 74.3 cc ($Vs + Vs = 50 + 24.3$). The change in volume from LL to PL is $98.6 - 74.3 = 24.3$ cc, and the percent change in volume is $100 \times 24.3 / 98.6 = 24.6\%$.

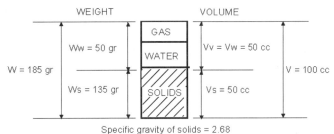

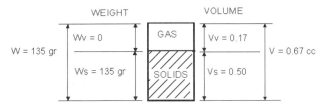

FIGURE 2.9 Block diagram.

REFERENCES

1. Sowers, G.F. 1979. *Introductory Soil Mechanics and Foundations: Geotechnical Engineerng*. Fourth Edition. New York: Macmillan Publishing.
2. Atterberg, A. 1911. "Uber die physikalishe Bodenuntersuchung und über die Plastizität der Tone (On the Investigation of the Physical Properties of Soils and on the Plasticity of Clays)." Int. Mitt. fr Bodenkunde. 1. 10-43.
3. ASTM. 1999. "Natural Building Stones; Soil and Rock; Geotextiles." 1999 Annual Book of ASTM Standards, 04.08, American Society for Testing and Materials, Philadelphia, Pa.
4. Skempton, A. W. 1953. "The Colloidal Activity of Clays," Proceedings of the Third International Conference on Soil Mechanics and Foundation Engineering. 1. Zurich.
5. Seed, H. B., Woodward, R. J., and Lundgren, R. 1962. "Prediction of Swelling Potential for Compacted Clays." Journal of the Soil Mechanics and Foundations Division. 88(SM3). 53-87.
6. USAEWES. 1960. The Unified Soil Classification System, Technical Memorandum No. 3-357, US Army Engineer Waterways Experiment Station, Vicksburg, Miss.
7. AASHTO. 1988. Manual of Subsurface Investigations, American Association of State Highway and Transportation Officials, Washington, D. C.
8. Soil Survey Staff. 1975. "Soil Taxonomy: A Basic System of Soil Classification for Making and Interpreting Soil Surveys." Agriculture Handbook No. 436, U. S. Department of Agriculture, Soil Conservation Service, GPO, Washington, D.C.
9. NAVFAC. DM 7.01 1982. Soil Mechanics, Design Manual 7.01, U. S. Department of the Navy, Naval Facilities Engineering Command, Alexandria, Va. September.
10. Deere, D. U. 1963. "Technical Description of Rock Cores for Engineering Purposes." Rock Mechanics and Engineering Geology. 1(1). 18.

CHAPTER 3
HYDRAULIC PROPERTIES OF SOIL AND ROCK

Engineering design problems often involve the water contained in soil and rock masses and the rate at which the water will move, or permeate, through the soil or rock under pressure. Typical problems include water supply and construction site drainage, both of which involve the quantity of water that can be retrieved from a well, and seepage pressures in or under a water-retaining structure such as a dam or a wall. All such problems require knowledge of the permeability of the soil or rock mass. This chapter deals only with the movement of water under pressure through continuous voids in a saturated soil or rock mass.

COEFFICIENT OF PERMEABILITY

Darcy's Law

The voids in a soil or rock mass are filled with fluids, which are normally a combination of a liquid and a gas. In geotechnical engineering practice, the liquid is invariably water and the gas is air.

According to Darcy's law, when pressure is applied to the water in the voids of a mass of saturated soil, the water will permeate through the soil with a velocity

$$v = ki \tag{3.1}$$

where v = Discharge velocity; that is, the rate at which water will flow in unit time through a unit gross cross-sectional area of soil perpendicular to the direction of flow
 k = Coefficient of permeability, which is a proportionality constant
 i = Hydraulic gradient, $\Delta h/L$, the head (pressure) loss over length of flow path.

Because geologists often deal with liquids other than water flowing through soil or rock, they generally refer to the proportionality constant k involving water flow as the *hydraulic conductivity*. Geotechnical engineers use the terms permeability and hydraulic conductivity interchangeably.

Darcy's law was developed for clean sands and assumes (a) that the relationship between the discharge velocity v and the hydraulic gradient i, is linear and (b) that the hydraulic gradient is sufficiently low that the flow is laminar. The discharge velocity in Equation 3.1 is based on the gross cross-sectional area of the soil. Because the void space occupies only a fraction of the cross-section, the real seepage velocity through the void spaces is greater than v.

The coefficient of permeability has the same units as velocity. In the English system, k is generally expressed as feet-per-minute or feet-per-day. In SI units, k is generally expressed as centimeters per second. The coefficient of permeability of soils is dependent on several factors, including grain-size distribution, pore-size distribution, void ratio, fluid viscosity, angularity of the grains, and the degree of saturation, as air bubbles impede water flow. The coefficient of permeability k varies widely for different soils. Typical values are given in Table 3.1.

TABLE 3.1. Typical values of coefficients of permeability for various soils [4] and [5]

Soil Type	k (cm/sec)	k (ft/min)	Drainage
Coarse sand and gravel	2 to 5×10^{-1}	4 to 10×10^{-1}	Good
Medium to coarse sand	1.5 to 2×10^{-1}	3 to 4×10^{-1}	Good
Medium sand	1 to 1.5×10^{-1}	2 to 3×10^{-1}	Good
Fine to medium sand	5 to 10×10^{-2}	1 to 2×10^{-1}	Good
Fine sand	2 to 5×10^{-2}	4 to 10×10^{-2}	Good
Very fine sand	5 to 20×10^{-3}	1 to 4×10^{-2}	Good
Silty sand	2 to 5×10^{-3}	4 to 10×10^{-3}	Good
Sandy silt	5 to 20×10^{-4}	1 to 4×10^{-3}	Good
Inorganic silts.	10^{-5}	2×10^{-5}	Poor
Organic silts, mixtures of sand silt and clay, glacial till, stratified clay deposits.	10^{-6}	2×10^{-6}	Poor
Impervious soils such as homogeneous clays below the zone of weathering.	10^{-9} to 10^{-7}	2×10^{-9} to 2×10^{-7}	Practically Impervious

Laboratory Tests for the Coefficient of Permeability

Two standardized laboratory test methods are used to determine the coefficient of permeability of soils, the *constant head test* and the *falling head test*.

Constant Head Permeability Test. The typical arrangement of the constant head permeability test is shown schematically in Figure 3.1.

In this laboratory procedure, a saturated soil sample is placed in a container of uniform cross-section and length L at a known and uniform density. Water is introduced at the upper end and maintained at a constant height h at a rate such that continuous flow is maintained from point A to point B through the saturated soil. The water exits at the lower end at the constant height h_2 so that a constant head loss, Δh, over a distance L is maintained. After a constant rate of flow is established, the water overflowing the lower end is collected and measured for a known time period.

The total quantity (volume) of water collected in a given time period may be expressed as

$$Q = kiAt \tag{3.2}$$

where Q = Volume of water collected
 k = Coefficient of permeability
 i = Hydraulic gradient, h/L
 A = Cross-sectional area of sample
 t = Duration of time for collection of water.

With all other factors known by measurement, Equation 3.2 can be rewritten for determination of the coefficient of permeability as

$$k = \frac{QL}{A \Delta h t} \tag{3.3}$$

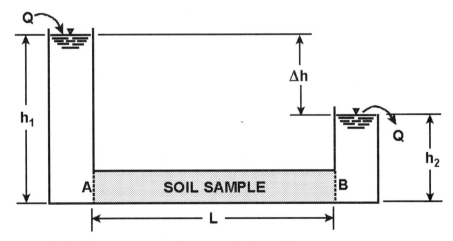

FIGURE 3.1 Schematic arrangement of the constant head permeability test.

Constant head permeability tests are particularly suitable for clean, coarse-grained soils having a high coefficient of permeability.

Falling Head Permeability Test. The typical arrangement of the falling head permeability test is shown schematically in Figure 3.2.

In this laboratory procedure, a saturated soil sample is placed in a container of uniform cross section and length L at a known and uniform density. Water is introduced at the upper end, above point A, in a standpipe at a rate such that continuous flow is maintained from point A to point B through the saturated soil. The water exits at the lower end at the constant height level so that a variable head loss over a distance L occurs over time.

After a continuous rate of flow is established at which the sample remains fully saturated, the level of water in the standpipe is recorded at time t_0 when the water in the standpipe is at level h_0 and the flow is continued until time t_1 when the level of water in the standpipe is at level h_1. The rate of flow of water through the soil at any time t can be given by

$$q = k\frac{h}{L}A = -a\frac{dh}{dt}$$

where a = cross-sectional area of the standpipe
 A = cross-sectional are of the soil sample
 L = length of the soil sample
 t = duration of time for the water level in the standpipe to drop from h_1 to h_2
 h_1 = water level in the standpipe at the start of time measurement
 h_2 = water level in the standpipe at the end of time measurement.

Rearrangement of the above equation gives

$$dt = \frac{aL}{Ak}\left(-\frac{dh}{h}\right)$$

When the left side is integrated with limits of time from t_0 to t_1 and the right side with limits of head difference from h_0 to h_1, the coefficient of permeability is calculated as

$$k = 2.303\,\frac{aL}{At}\log_{10}\frac{h_0}{h_1} \qquad (3.4)$$

The falling head test is appropriate in the case of fine-grained soils with low permeability.

Equivalent Permeability of Stratified Deposits

Many soil deposits contain layers that differ in grain size and permeability. Within each layer in natural soil deposits, the coefficient of permeability may differ between the horizontal and vertical directions with a ratio of horizontal to vertical ranging from as little as one or two to almost ten.

Equivalent Horizontal Permeability. When water moves horizontally through several layers of different permeabilities, the equivalent horizontal permeability may be almost as great as the permeability of the most permeable layer.

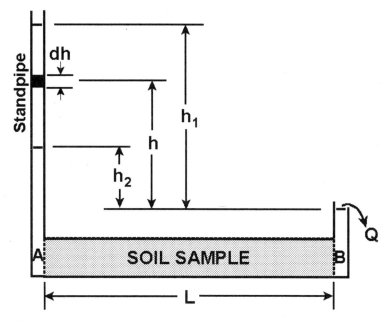

FIGURE 3.2 Schematic arrangement of the falling head permeability test.

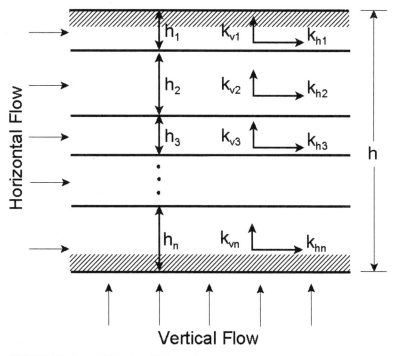

FIGURE 3.3 Permeability of stratified deposits.

Referring to Figure 3.3, and applying Darcy's law, Equation 3.1,

$$v = k_{h(eq)} \times i_{eq}; \; v_1 = k_{h1} \times i_1; \; v_2 = k_{h2} \times i_2; \; v_3 = k_{h3} \times i_3; \; \ldots; \; v_n = k_{hn} \times i_n$$

where v = average discharge velocity
$v_1, v_2, v_3, \ldots, v_n$ = discharge velocities of the individual layers denoted by subscripts
$k_1, k_1, k_1, \ldots, k_1,$ = coefficients of permeability of the individual layers.

Then, noting that the hydraulic gradient is the same for all layers, that is, $i_{eq} = i_1 = i_2 = i_3 = \ldots = i_n$, the *equivalent horizontal permeability* becomes

$$k_{h(eq)} = \frac{1}{h}\left(k_{h1}h_1 + k_{h2}h_2 + k_{h3}h_3 + \ldots + k_{hn}h_n\right) \tag{3.5}$$

Equivalent Vertical Permeability. When water moves vertically through several layers of different permeabilities, the equivalent vertical permeability may be almost as small as the permeability of the least permeable layer. Referring to Figure 3.3, the velocity of flow is the same for all layers, that is, $v_{eq} = v_1 = v_2 = v_3 = \ldots = v_n$. The total head loss h is equal to the sum of the head loss in each layer, that is, $h = h_1 + h_2 + h_3 + \ldots + h_n$. Then, substituting in Darcy's law, Equation 3.1, the *equivalent vertical permeability* becomes

$$k_{v(eq)} = \frac{h}{\left(\dfrac{h_1}{k_{v1}}\right) + \left(\dfrac{h_2}{k_{v2}}\right) + \left(\dfrac{h_3}{k_{v3}}\right) + \ldots + \left(\dfrac{h_n}{k_{vn}}\right)} \tag{3.6}$$

Empirical Estimates for Coefficient of Permeability

Generally, the coefficient of permeability increases with the size of voids, which in turn increases with increasing grain size and with uniformity of the distribution. No simple relationships, however, have been found between permeability and grain size except in the case of fairly coarse soils with rounded grains.

Allen Hazen [1], in studies of the permeability of loose filter sands, found that

$$k = CD_{10}^2 \tag{3.7}$$

where k is expressed in centimeters/second
C varies from 0.4 to 1.5 (usually approximated as 1.0)
D_{10} is the effective grain diameter corresponding to 10 percent finer, expressed in millimeters, as defined in Chapter 2. The Hazen formula is generally limited to sands with values of D_{10} between 0.1 and 3 mm and a coefficient of uniformity C_u of less than 5.

Later experimenters developed additional correlations between k and D_{10} for other grain-size conditions. Some of these correlations have been summarized [2] in the graph of Figure 3.4, which includes Burmister's laboratory tests of compacted samples at Columbia University and Mansur's field pumping test and laboratory permeameter correlations for sands in the Mississippi and Arkansas River valleys.

Permeability of Rock

The permeability of sound rock is of little more than academic interest to the civil engineer in the design of structures and facilities. Virtually every near-surface deposit of rock contains cracks, joints, bedding planes, or solution cavities. The influence of these features on the permeability of the rock mass is beyond the scope of this book. It can be stated, however, that the permeability can vary over a wide range within short distances and is best determined by field tests.

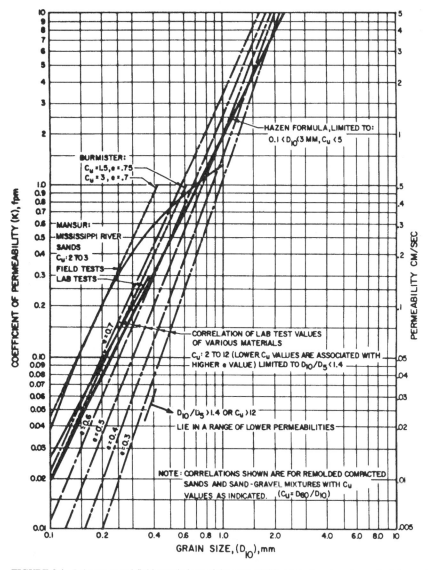

FIGURE 3.4 Laboratory and field correlations of the permeability of sands and sand-gravel mixtures with effective grain size, D10 [2]

Field Tests for Permeability by Pumping from Wells

The coefficient of permeability to be used in estimating the pumping rate from a well cannot be determined accurately by laboratory tests because, within the zone of influence of the well, minor variations occur in the grain size and density of the soil, in the direction of flow, and in the ratio of horizontal to vertical permeability in each layer. Therefore, the most effective manner of determining the accurate field permeability is a full-scale pumping test from a well.

Theoretical calculations for the pumping rate involve not only the stratigraphic geometry of the well system but the rate of flow q and the coefficient of permeability k. With two unknowns, it is necessary either (a) to make two tests at the same well using two different pumping rates q and solve the two equations simultaneously, or (b) to solve for one variable by means of a field pumping test and then use that value to estimate the value of the other rate for similar wells in the same or similar soil or rock deposits. Two geometry conditions exist for circular flow to a well, artesian flow from a confined aquifer and gravity flow. In addition, there are two penetration conditions into the aquifer, fully penetrating and partially penetrating.

Well Pumping Test. A test well is made with a length of standard perforated casing through the aquifer below the static water level. Several observation wells, consisting of piezometers, are placed in one or more radial lines and at various radial distances from the test well. Pumping of the test well continues until a steady state is reached in the water levels in the test well and in the observation wells. A single test well draws from a cylinder of soil with a radius of R. The rate of flow of ground water into the well equals the rate of discharge due to pumping. The direction(s) of flow of groundwater toward the well and the governing coefficient of permeability depend on the well geometry.

Artesian Flow—Fully Penetrating Well. An artesian well is one that taps an aquifer confined between two impervious layers in which hydrostatic pressure exists against the upper confining layer. In the case of a fully penetrating well, Figure 3.5, the flow of groundwater toward the well is essentially horizontal.

Using field measurements, the coefficient of permeability can be calculated from the *Theim equation*:

$$k = \frac{2.303q}{2\pi D(h_2 - h_1)} \log_{10} \frac{r_2}{r_1} \tag{3.8}$$

where
- k = Effective coefficient of permeability
- q = Rate of steady-state flow from well
- D = Thickness of aquifer
- r_1 and r_2 = Horizontal distances from center of well to observation wells 1 and 2
- h_1 and h_2 = Heights of water in observation wells 1 and 2 above datum
- r_w = Radius of test well
- h_w = Height of water in test well above datum
- d_w, d_1, and d_2 = Drawdown at the pump well, well 1, and well 2, respectively.

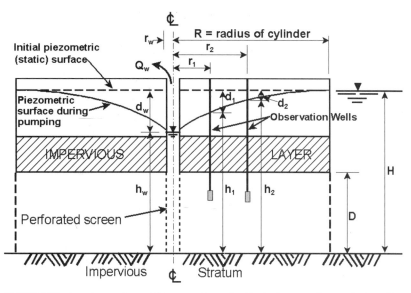

FIGURE 3.5 Pumping test from a fully penetrating well with circular source in a confined aquifer (artesian flow).

Artesian Flow—Partially Penetrating Well. In the case of a partially penetrating artesian well, Figure 3.6, the flow of groundwater toward the well is essentially hemispherical, ranging from horizontal near the top of the aquifer to nearly vertical at the bottom of the well.

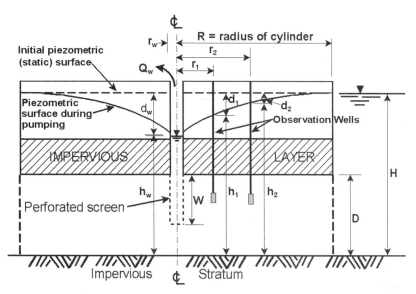

FIGURE 3.6 Pumping test from a partially penetrating well with circular source in a confined aquifer (artesian flow).

Using field measurements, the coefficient of permeability can be calculated from the *Kozeny equation*

$$k = \frac{2.303q}{2\pi D(h_2 - h_1)G} \log_{10} \frac{r_2}{r_1} \tag{3.9}$$

where the approximate value of G can be computed from

$$G = \frac{W}{D}\left(1 + 7\sqrt{\frac{r_W}{2W}} \cos\frac{\pi W}{2D}\right) \tag{3.10}$$

Gravity Flow—Fully Penetrating Well. A gravity, or ordinary, well is one in which water is drawn from below the natural, or static, groundwater table. Unlike the artesian well, there is no excess hydrostatic pressure against a confining layer to help bring the water to the surface. In the case of a fully penetrating well, Figure 3.7, the flow of groundwater toward the well is essentially horizontal near the bottom of the aquifer but is inclined downward along the drawdown curve in the upper part of the aquifer.

Using field measurements, the coefficient of permeability can be calculated from the *Dupuit equation*

$$k = \frac{2.303q}{\pi\left(h_2^2 - h_1^2\right)} \log_{10} \frac{r_2}{r_1} \tag{3.11}$$

Gravity Flow—Partially Penetrating Well. In the case of a partially penetrating gravity well, Figure 3.8, the flow of groundwater toward the well is nearly spherical, ranging from being inclined along the drawdown curve near the top of the well to being horizontal at the center of the well to being nearly vertical at the bottom of the well.

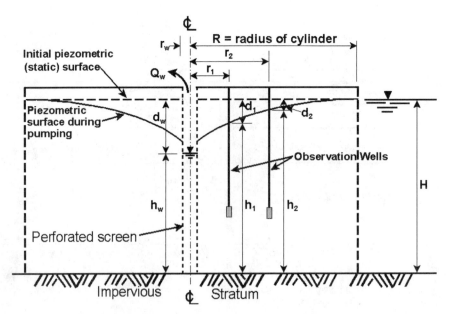

FIGURE 3.7 Pumping test from a fully penetrating well with circular source and gravity flow.

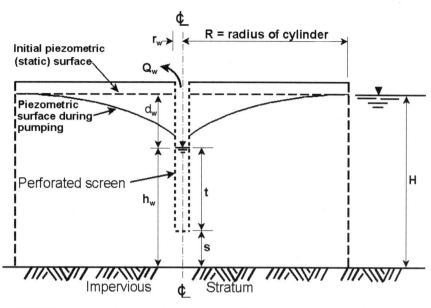

FIGURE 3.8 Pumping test from a partially penetrating well with circular source and gravity flow.

For the case where $(H - s)/H \geq 0.85$, that is, the pump well penetrates the aquifer by at least 85 percent of its depth, Equation 3.11 is applicable. Where $(H - s)/H < 0.85$, more exact solutions are available using graphs for constants [3]. For a reasonable approximation using field measurements, the coefficient of permeability can be calculated from [4]

$$k = \frac{2.303q}{\pi C\left[(H - s)^2 - t^2\right]} \log_{10} \frac{R}{r_w} \qquad (3.12)$$

where

$$C = 1 + \left(0.30 + \frac{10 r_w}{H}\right) \sin \frac{1.8s}{H} \qquad (3.13)$$

In Equation 3.12, the factor C of Equation 3.13 may be ignored for a large range of practical values for the other terms because it is equal to slightly more than 1.00. The radius of influence of the well R should be taken as a very large value, such as $20H$ or more. This will yield an estimate of k within an order of magnitude.

EFFECTIVE STRESS AND SEEPAGE PRESSURE

In a saturated soil mass, the voids between the grains are completely filled with water. At any depth within this mass, the vertical pressure is a combination of water pressure and grain-to-grain pressure. When the water flows upward or downward through the soil mass, both the water pressure and the grain pressure change.

Principle of Effective Stress

The container shown in Figure 3.9 [5, 6], is filled with a clean granular soil from its bottom to a height H_w. Water fills the container from its bottom to the total height, $H_w + H_o$, so that the soil is fully saturated. The standpipe connected to the bottom of the container is filled with water to the level of the top of the container so that there is no flow of water in or out.

On plane ab at depth ($H_o + z$) below the top of the container, the total vertical pressure p due to water plus soil is

$$p = H_o \cdot \gamma_w + z \cdot \gamma_{sat} \qquad (3.14)$$

where γ_w = Unit weight of water
γ_{sat} = Unit weight of the saturated soil.

The water above ab extends in continuous voids to the height z and as a continuous mass above the soil to the height H_o. Therefore, the pore water pressure u_w at the surface ab is

$$u_w = (H_o + z)\gamma_w \qquad (3.15)$$

The *Terzaghi principle of effective stress* states that the total vertical stress on a plane consists of

$$p = \bar{p} + u_w \qquad (3.16)$$

where $\bar{p}$ is the effective vertical pressure. Therefore,

$$\bar{p} = z(\gamma_{sat} - \gamma_w) \qquad (3.17)$$

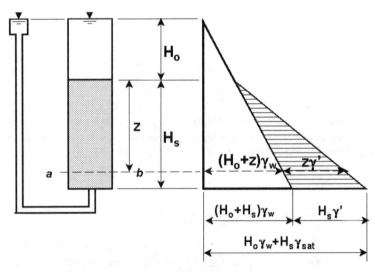

FIGURE 3.9 Diagram illustrating neutral flow condition. [5]

In Equation 3.17, the quantity $(\gamma_{sat} - \gamma_w)$ is known as the *submerged unit weight*, γ'. Therefore, for the no-flow conditions of Figure 3.9, the effective stress on plane *ab* is

$$\bar{p} = z\gamma' \tag{3.18}$$

which shows that the effective pressure on plane *ab* is independent of the height of water H_o above the submerged soil or rock.

In Figure 3.9, the width of the unshaded area at any depth represents the distribution of pore water pressure, Equation 3.15. The width of the shaded area at any depth represents the distribution of the effective pressure, Equations 3.17 and 3.18, and the total pressure at any depth is the sum of the two, Equation 3.16.

Seepage Pressures and the Critical Hydraulic Gradient

Downward Flow. When water flows *downward* through the soil, then Equations 3.15, 3.17, and 3.18 are no longer applicable.

If the water surface is maintained at the top of the container but the top of the reservoir is lowered, as in Figure 3.10, and a steady state of flow is established downward, then the pore water pressure on the plane *ab* is

$$u_w = \frac{z}{H_s}\left(H_o + H_s - h\right)\gamma_w \tag{3.19}$$

Since the total pressure on plane *ab* remains the same, the effective pressure is increased by the value

$$\bar{p} = \frac{z}{H_s} h\gamma_w \tag{3.20}$$

which is known as the *seepage pressure*, and is the result of frictional downdrag of the

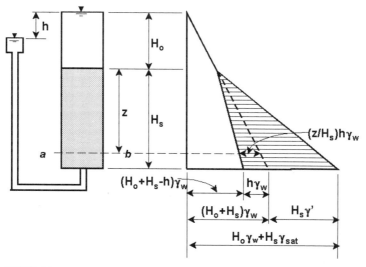

FIGURE 3.10 Diagram illustrating downward flow (seepage pressure) condition. [5]

flowing water on the grains. If the hydraulic gradient i at depth z is expressed as h/H_s, then the effective pressure at depth z due to *downward* flowing water is

$$\bar{p} = z\gamma' + iz\gamma_w \qquad (3.21)$$

Upward Flow. When water flows *upward* through the soil, Equations 3.15, 3.17, and 3.18 are again no longer applicable.

If the water surface is maintained at the top of the container but the top of the reservoir is raised, as in Figure 3.11, and a steady state of flow is established upward, then the pore water pressure on the plane *ab* is

$$u_w = \frac{z}{H_s}(H_o + H_s + h)\gamma_w \qquad (3.22)$$

Since the total pressure on plane *ab* remains the same, the effective vertical pressure is *decreased* by the seepage pressure of Equation 3.20, which results from the frictional updrag of the flowing water on the grains. If the hydraulic gradient i at depth z is expressed as h/H_s, then the effective pressure at depth z is reduced to

$$\bar{p} = z\gamma' - iz\gamma_w \qquad (3.23)$$

By increasing the hydraulic gradient, $i = h/H_s$ at depth z so that $(z\gamma' - i_c z\gamma_w) = 0$, we can define

$$i_c = \frac{\gamma'}{\gamma_w} \qquad (3.24)$$

as the *critical hydraulic gradient*, at which the effective pressure becomes zero and the granular soil can no longer support a load on its surface.

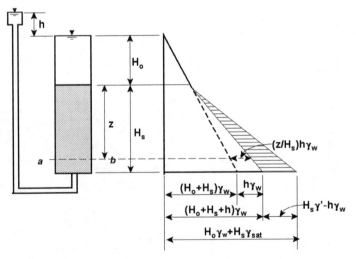

FIGURE 3.11 Diagram illustrating upward flow (seepage pressure) condition. [5]

SEEPAGE OF WATER THROUGH SOILS

The pattern of the water flow through a natural soil deposit is not uniform because of the erratic nature of permeability from point-to-point and in different directions. In the simplest of such cases, calculation of the ground water flow is generally made by use of graphs known as *flow nets*. The concept of the flow net is based on *Laplace's equation of continuity* for two-dimensional flow,

$$k_x \frac{\partial^2 h}{\partial x^2} + k_z \frac{\partial^2 h}{\partial z^2} = 0 \tag{3.25}$$

For all cases, the uppermost flow surface of the water either (a) is confined by an impervious surface, against which the water exerts pore water pressure, or (b) is a free surface, on which the pressure in the pore water is zero.

Flow Net in Isotropic Soil

Consider the case where an impervious surface exists. If water stands on each side of an impervious barrier, the seepage line is horizontal. A droplet of water entering the ground surface below the upstream side of the sheet pile wall in Figure 3.12 follows a path through the soil known as a *flow line*.

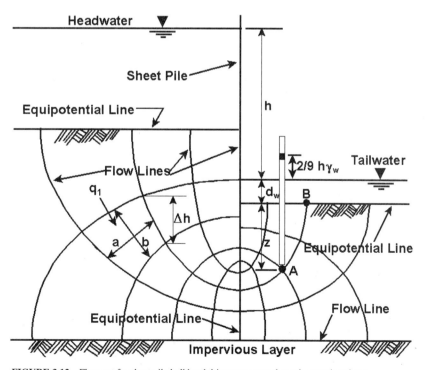

FIGURE 3.12 Flow net for sheet pile bulkhead driven part way through a pervious layer.

The flow line is divided into a number of points of equal head loss, Δh, so that $\Sigma \Delta h = h$, the total head loss from the upstream to the downstream side.

The Laplace equation, Equation 3.25, represents two orthogonal families of curves—that is, curves meeting at 90°—consisting of ellipses and hyperbolas. The ellipses represent the flow lines and the intersecting lines represent the *equipotential lines*. At all points on an equipotential line, as seen in Figure 3.13, the remaining head is the same so that the water in a standpipe inserted at any point on the line will rise to the same level.

If several standpipes are inserted in the next equipotential line, the hydrostatic level will be the same at all points, but lower than the previous equipotential line by Δh. It should be observed, however, that the equipotential lines do not join points of equal pressure. The pore water pressure at any point, such as (x, y) in Figure 3.13, is equal to the head of water in the tube times the unit weight of water, and is not equal to the pressure at another point on the same equipotential line.

A combination of flow line and equipotential lines is called a flow net. In a practical flow net, there are several flow lines, chosen in such a way that the quantity of water q_1 flowing between any two flow lines (flow channel) is the same as that flowing between any other two flow lines. Thus, the net quantity of water flowing through the soil q_T is equal to q_1 times the number of flow channels in the net N_f so that $q_T = N_f \times q_1$. In such a flow net the total drop in head h is equal to the number of equipotential intervals N_d so that $h = \Delta h\, N_d$.

Consider the rectangle shown in Figure 3.12, which has a distance between flow lines equal to a and a distance between equipotential lines equal to b. Applying Darcy's law for a unit's thickness, the flow through this channel is

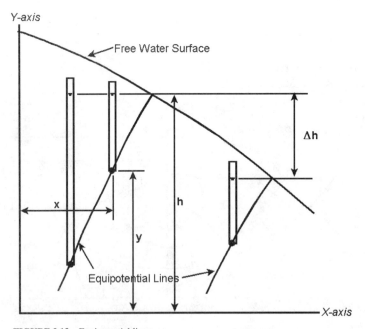

FIGURE 3.13 Equipotential lines.

$$q_1 = kia = k\frac{\Delta h}{b} a$$

If a flow net is drawn to consist of a series of rectangles so that the sides a and b are equal, forming squares of each rectangle, then the resulting flow net will be a true one so that

$$q_T = kh \frac{N_f}{N_d} \quad (3.26)$$

is the total quantity of water that will flow in unit time through an isotropic soil deposit.

Drawing a Flow Net

Flow nets are drawn by eye and are reasonably accurate for isotropic soils. Familiarity with the flow nets for several simple conditions is helpful. The following steps should be followed:

1. Evaluate the boundary conditions to define all definite flow and equipotential lines.
   (a) Flow lines enter and leave pervious surfaces at right angles.
   (b) Equipotential lines enter and leave impervious surfaces at right angles.
2. Flow lines and equipotential lines are orthogonal sets; that is, they intersect at right angles.
3. The ratio of the distance between adjacent flow lines and adjacent equipotential lines is constant. Usually this ratio is one in which squares are obtained.
4. Start by guessing two or three flow lines reflecting ellipses that have been distorted by the shape of the bounding conditions. These should be fairly smooth curves.
5. Sketch in equipotential lines (distorted hyperbolas) so that they form squares with the flow lines and intersect the flow lines at right angles.
6. Check for squares. Can a circle be drawn in each one?
7. The two extreme squares at each end of the flow net will not be complete.

Flow Net in Anisotropic Soil

For anisotropic soils, the horizontal and vertical permeabilities in Laplace's Equation 3.25, are not equal. The equation, however, is valid if $x' = x \cdot \sqrt{k_z/k_y}$ is substituted for x. To construct the flow net,

1. Use a horizontal scale (x-axis) such that the horizontal scale is $\sqrt{k_z/k_x}$ times the vertical scale.
2. Draw the flow net in the normal manner, with flow lines intersecting equipotential lines at right angles and the elements forming approximate squares.
3. Redraw the flow net at the normal scale for both axes, recognizing that the lines no longer intersect at right angles.

4. Modify Equation 3.26 to calculate the rate of seepage per unit width as

$$q = \sqrt{k_x \cdot k_z}\, H \frac{N_f}{N_d} \tag{3.27}$$

Seepage Line—Phreatic Surface

A water surface at zero pore water pressure, is called the *free surface* or the *phreatic surface*. If the seepage line does not encounter an impervious barrier, the change in potential head along the phreatic surface Δh is the actual vertical distance from one point on the free surface to another.

Before a flow net for water flowing through the soil can be drawn, the seepage line must be determined. Using assumptions suggested by Dupuit [7], it can be shown that the seepage line for a homogeneous, isotropic soil is a parabola.

The free surface must come to the discharge point at some distance above the tailwater surface or above an impervious layer. When the discharge face is a slope of less than 90°, the seepage line is tangent to the discharge face, as shown in Figure 3.14.

The point of tangency is

$$a = \frac{d}{\cos\beta} - \sqrt{\frac{d^2}{\cos^2\beta} - \frac{h^2}{\sin^2\beta}} \tag{3.28}$$

On the upstream face of an earth dam, which is an equipotential line, the water enters the embankment at a right angle to the face. This flow line, however, must turn immediately to form the square of a flow net. A short curve perpendicular to the slope at the entrance to the parabola corrects the seepage line in that portion. As determined empirically by Casagrande [8], the true parabola intersects the upstream water surface at a point $m/3$, as shown in Figure 3.15, where m is the horizontal distance from the upstream toe to the intersection of the impounded water with the upstream slope.

The remainder of the flow net must follow the same steps and restrictions described above for a flow net involving an impervious barrier. Where the base of the embankment is not an impervious barrier, the flow net continues below the embankment in the usual manner. For the normal case, in which the horizontal and vertical permeabilities are not equal, the entire cross-section is modified as described above for flow through anisotropic soil, and the quantity of flow determined by Equation 3.27.

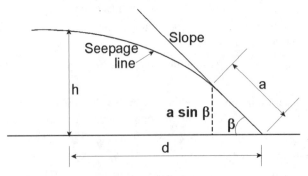

FIGURE 3.14 Seepage line tangent to discharge slope.

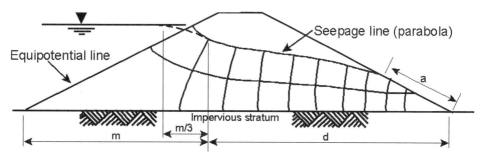

FIGURE 3.15 Flow net for seepage through a dam of homogeneous, isotropic material.

Heaving of Soil at Exit Point

In any flow net, the head drop between successive equipotential lines is $\Delta h = h / N_d$. For example, in Figure 3.12 the number of drops is 9. Along the equipotential line containing point A, the total head loss $7\Delta h = 7h/9$. The pressure in the pore water at point A is

$$u_A = \gamma_w \left(z_A + d_w + \frac{2}{9}h \right)$$

The *excess pore water pressure* (above hydrostatic) at A with respect to the tailwater level is the pressure available at A to drive the water the remaining distance to B, and is equal to $2/9\, h\, \gamma_w$, as shown in Figure 3.12.

The upward seepage force is greatest at the point of exit nearest a structure and may cause heave or boiling of the soil when the hydraulic gradient is too large. For sheet pile structures such as that shown in Figure 3.12, Terzaghi [9] made model tests and concluded that heave generally occurs within a distance $D/2$ from the sheet piles, where D is the depth of embedment into the permeable layer. Therefore, the stability of a zone measuring D deep by $D/2$ wide at the exit point next to the sheet pile is of concern. The factor of safety FS against heave is given by

$$FS = \frac{W'}{U} \qquad (3.29)$$

where W' = Submerged weight of soil in the heave zone per unit width of sheet pile, and
U = Uplift force due to seepage on the same volume of soil.

W' is computed by

$$W' = D\frac{D}{2}(\gamma_{sat} - \gamma_w) = \frac{1}{2}D^2\gamma' \qquad (3.30)$$

and U is given by

$$U = \frac{1}{2}D^2 \times i_{AV}\gamma_w \qquad (3.31)$$

where i_{av} = average hydraulic gradient in the block of soil.

Substituting the values of W' and U in Equation 3.29, we can write

$$FS = \frac{\gamma'}{i_{AV}\gamma_w} \tag{3.32}$$

The average hydraulic gradient can be determined from the flow net. It is the average of the excess pore water pressure with respect to the tailwater level, as defined above, at the two ends of the block $D/2$ by D.

EXAMPLE PROBLEMS

Constant Head Permeability Test

3.1 A constant head permeability test was made on a sample of sand whose cross-sectional area is 800 cm^2 and whose length is 30 cm. After steady flow is established at a constant head difference of 50 cm, 35 cc of water is collected at the discharge end every 5 minutes. Calculate the value of the coefficient of permeability.

SOLUTION:
Using Equation 3.3,

$$k = \frac{QL}{A\Delta ht} = \frac{35 \times 30}{800 \times 50 \times 5} = 0.00525 \text{ cm/sec}$$

$$k = \sim 5 \, (10^{-3} \text{ cm/sec}).$$

Falling Head Permeability Test

3.2 A falling head permeability test was made on a sample of clayey silt whose cross-sectional area is 50 cm^2 and whose length is 10 cm. The cross-sectional area of the inside of the standpipe was 0.03 cm^2. After steady flow was established, the head on the sample decreased from 75 cm to 50 cm over 30 minutes. Calculate the value of the coefficient of permeability.

ANSWER:
Using Equation 3.4,

$$k = 2.303 \frac{aL}{At} \log_{10} \frac{h_0}{h_1} = 2.303 \times \frac{0.03 \times 10}{50 \times 30} \log_{10} \frac{75}{50}$$

$$k = 0.000081 \text{ cm/sec} = \sim 8 \, (10^{-5} \text{ cm/sec})$$

Equivalent Horizontal Permeability

3.3 Find the equivalent coefficient of permeability for flow in the horizontal direction for a soil deposit having three layers with the following characteristics:

COEFFICIENT OF PERMEABILITY, FT/MIN

Layer No.	Thickness, ft	In Horizontal Direction	In Vertical Direction
1	25	1.5×10^{-2}	3×10^{-3}
2	18	2×10^{-1}	5×10^{-2}
3	9	3×10^{-4}	10^{-4}

SOLUTION:
From Equation 3.5,

$$k_{h(eq)} = \frac{1}{52}(0.015 \times 25) + (0.2 \times 18) + (0.0003 \times 9) = 0.076 \text{ ft/min}$$

$$k_{h(eq)} = \sim 8 \times 10^{-2} \text{ ft/min}$$

Equivalent Vertical Permeability

3.4 Find the equivalent coefficient of permeability for flow in the vertical direction for a soil deposit having the layering and characteristics described in Problem 3.3.

SOLUTION:
From Equation 3.6,

$$k_{v(eq)} = \frac{52}{\left(\frac{25}{0.003}\right) + \left(\frac{18}{0.05}\right) + \left(\frac{9}{0.0001}\right)}$$

$$k_{v(eq)} = 0.00053 \text{ ft/min} = \sim 5 \times 10^{-4} \text{ ft/min}$$

Estimating the Coefficient of Permeability

3.5 A clean, cohesionless sand with rounded grains has an effective grain size, D_{10}, equal to 0.17 mm. Estimate the coefficient of permeability using the Hazen formula.

SOLUTION:
Using Equation 3.7, $k = CD_{10}^2 = 1.0 \times (0.17)^2 = 0.0289$ cm/sec $= \sim 3 \times 10^{-2}$ cm/sec.

Well Pumping Test—Fully Penetrating Artesian Flow

3.6 Refer to Figure 3.5. A field pumping test was made in a sand deposit overlaid by a clay deposit. The static groundwater table is 10 ft below ground surface. The bottom of the clay layer is at a depth of 50 ft. The sand deposit is underlaid by an impervious clay at a depth of 90 ft. The test well was drilled to fully penetrate the sand layer. Two observation wells were placed in a radial direction from the test well. Observation well Number 1 was 15 ft and Number 2 100 ft from the test well. When a steady state was reached at a pumping rate of 5 gal/min, the drawdown at the outer well (Number 2) was 1.2 ft and at the inner well (Number 1) it was 3.0 ft. What was the coefficient of permeability of the sand?

SOLUTION:
Using Figure 3.5, the following values were measured or calculated:

$q = 5$ gpm $= 0.67$ cfm
$r_1 = 15$ ft
$r_2 = 100$ ft
$H = 90 - 10 = 80$ ft
$h_1 = 80 - 3 = 77$ ft
$h_2 = 80 - 1.2 = 78.8$ ft
$D = 90 - 50 = 40$ ft.

Then using Equation 3.8,

$$k = \frac{2.303 \times 0.67}{2\pi \cdot 40(78.8 - 77)} \log_{10} \frac{100}{15} = 0.0028 \text{ ft/min}$$

$k = \sim 3 \times 10^{-3}$ ft/min.

Well Pumping Test—Partially Penetrating Artesian Flow

3.7 Refer to Figure 3.6. If the test well for the pumping test of Problem 3.6 had a diameter of 1.0 ft ($r_w = 0.5$ ft) and was drilled to a depth of only 75 ft, calculate the steady state pumping rate from this well. (Hint: Use Equation 3.10 and then rearrange Equation 3.9 to yield quantity of flow q using k from Problem 3.6).

SOLUTION:
Using Equation 3.10,

$$G = \frac{25}{40}\left[1 + 7\sqrt{\frac{0.5}{2 \times 25}} \cos \frac{3.14 \times 25}{2 \times 40}\right] = 1.0624$$

Equation 3.9 gives:

$$q = \frac{2 \times 3.14 \times 0.0028 \times 40 \times (78.8 - 77) \times 1.0624}{2.303 \log_{10}(100/15)} = 0.709 \text{ cfm}$$

$q = 5.3$ gal/min.

Well Pumping Test—Fully Penetrating Gravity Flow

3.8 Refer to Figure 3.7. A field pumping test was made in a sand deposit. The static groundwater table is 10 ft below the ground surface. The sand deposit is underlaid by an impervious clay at a depth of 90 ft. The test well was drilled to fully penetrate the sand layer. Two observation wells were placed in a radial direction from the test well. Observation well Number 1 was at 15 ft and Number 2 at 100 ft from the test well. When a steady state was reached at a pumping rate of 5 gal/min, the drawdown at the outer well (Number 2) was 1.2 ft and at the inner well (Number 1) was 3.0 ft. What was the coefficient of permeability of the sand?

SOLUTION:
Using Figure 3.7, the following values were measured or calculated:

where $q = 5$ gpm $= 0.67$ cfm
$r_1 = 15$ ft
$r_2 = 100$ ft
$H = 90 - 10 = 80$ ft
$h_1 = 80 - 3 = 77$ ft
$h_2 = 80 - 1.2 = 78.8$ ft.

Then using Equation 3.11,

$$k = \frac{2.303 \times 0.67}{\pi (78.8^2 - 77^2)} \log_{10} \frac{100}{15} = 0.0014 \text{ ft/min}$$

$k = \sim 1 \times 10^{-3}$ ft/min

Well Pumping Test—Partially Penetrating Gravity Flow

3.9 Refer to Figure 3.8. Assume the test well for the pumping test of Problem 3.6 had a diameter of 1.0 ft ($r_w = 0.5$ ft), was drilled to a depth of only 75 ft, and had a drawdown of 15 ft. Calculate the steady-state pumping rate from this well. (Hint: Rearrange Equation 3.11 to yield quantity of flow q and use k from Problem 3.8).

SOLUTION:

$h_w = 90 - 10 - 15 = 65$ ft. Since $s = 15$ ft, $t = 50$ ft.

Then, $q = \dfrac{\pi \times 0.0014 \times \left[(80 - 15)^2 - 50^2\right]}{2.303 \times \log_{10} (2000/0.5)} = 0.29$ ft^3/min

$q = 2.2$ gal/min.

Effective Stress

3.10 A soil profile consists of a dry sand from the surface to a depth of 8 m, underlaid by a saturated clay to a depth of 24 m, which in turn is underlaid by an impervious stratum. The groundwater table is at the surface of the clay layer. The dry unit weight of the sand is 16.0 kN/m³ and the saturated unit weight of the clay is 19.0 kN/m³. Calculate the total pressure, pore water pressure, and effective pressure at (a) the center of the sand layer, (b) the bottom of the sand layer, (c) the center of the clay layer, and (d) the bottom of the clay layer. The unit weight of water is 9.81 kN/m³.

SOLUTION:
See Table 3.2

Seepage Pressures and Critical Hydraulic Gradient

3.11 What is the critical hydraulic gradient for a medium-to-coarse sand having the following characteristics: permeability, $k = 2 \times 10^{-1}$ cm/sec, specific gravity of grains = 2.65, and void ratio, $e = 0.60$?

SOLUTION:
Using Equation 3.24, it is necessary to determine the buoyed unit weight.
 Referring to Chapter 2 and the unit cube shown in Figure 3.16 and assuming the dry weight of a volume of the soil is 100 lbs., then the volume of soil Vs must be equal to 100 / (2.65 × 62.4) = 0.605 ft³. For a void ratio of 0.6, the total volume must be $V = Vs + 0.6\, Vs = 1.6\, Vs = 1.6 \times 0.606 = 0.968$ ft³. The volume of water is then $V - Vs = 0.968$

TABLE 3.2. Solutions to Problem 3.10

	Total Pressure, eq. 3.14, kN/m².	Pore Water Pressure, eq. 3.15, kN/m².	Effective Pressure, eq. 3.16, kN/m².
(a) Center of sand	4 × 16 = 64x	0	4 × 16 = 64
(b) Bottom of sand	8 × 16 = 128	0	8 × 16 = 128
(c) Center of clay	128 + 8 × 19 = 280	8 × 9.81 = 78.5	280 − 78.5 = 201.5
(d) Bottom of clay	128 + 16 × 19 = 432	16 × 9.81 = 157	432 − 157 = 275

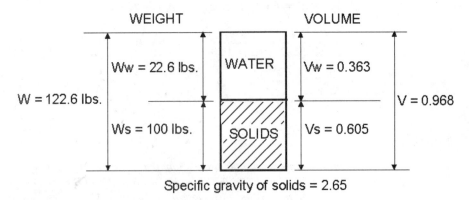

FIGURE 3.16 Unit cube to accompany Example Problem 3.11.

$-0.605 = 0.363$ ft³. The water weighs $0.363 \times 62.4 = 22.6$ lbs. The total weight of soil plus water is then $100 + 22.6 = 122.6$ lbs. The saturated unit weight is $122.6 / 0.968 = 126.7$ lb/ft³. The critical hydraulic gradient from Equation 3.23 is

$$i_c = \frac{\gamma'}{\gamma_w} = \frac{126.7 - 62.4}{62.4} = 1.03$$

Flow Net in Isotropic Soil

3.12 The concrete dam 25 m long shown in Figure 3.17 is embedded 1 m into the ground surface and has a sheet pile wall 5 m deep at its heel. The headwater is 7 m deep and the tailwater is at ground surface. The permeability of the soil is $k = 20 \times 10^{-4}$ cm/sec both vertically and horizontally.

a. What is the quantity of discharge per meter of width?

b. What is the uplift pressure distribution on the bottom of the dam?

SOLUTION:

a. As shown in Figure 3.16, there are $N_f = 4$ flow paths and $N_d = 13$ equipotential drops. Thus, from Equation 3.26,

$q = (N_f / N_d) \times k \times h = (4/13) \times 0.0002 \times 100 \times 7$

$q = 0.043$ m³ per sec. per meter width.

b. The water pressure on the ground surface at the bottom of the upstream end is 7 m times the unit weight of water, 9.81 kN/m³, which equals 68.7 kN/m². At the heel of

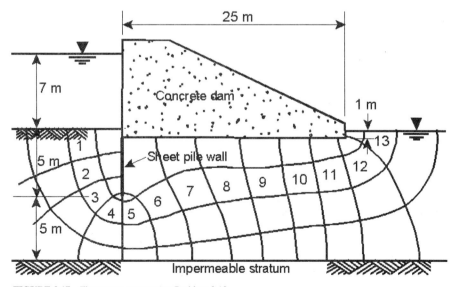

FIGURE 3.17 Flow net to accompany Problem 3.12.

the dam, the head loss is about 5.5 / 13 of the head and about 7.5 / 13 of the head remains. Therefore, the pressure at the heel of the dam is $p = 9.81 \times 7 \times 7.5/13 = 39.6$ kN/m². At the toe of the dam, about 1.5 / 13 of the head remains and $p = 9.81 \times 7 \times 1.5 / 13 = 7.92$ kN/m². The pressure distribution on the rest of the bottom of the dam is a linear distribution between these two end-pressures.

Heaving of Soil at Exit Point

3.13 *Two sheet pile walls, 20 ft apart, are each driven 16 ft below the original surface of a sand deposit and the sand removed for a depth of 10 ft, as shown in Figure 3.18.*

The depth of water outside the sheet pile walls is 10 ft. The inside water level is at the bottom of the excavation, 20 ft below the outside water level. The saturated unit weight of the soil in the permeable layer is 120 lb/ft³ and the permeability of the soil is $k = 20 \times 10^{-4}$ cm/sec (40×10^{-4} ft/min) both vertically and horizontally.

a. What is the quantity of seepage per foot of length of the cofferdam?

b. What is the factor of safety against upheaval of the bottom?

SOLUTION:

a. The cofferdam cross section is symmetrical. Therefore, we can deal with one-half of the section. There are 3 flow lines and 7 potential drops in each half. From Equation 3.26, $q = (N_f / N_d) \times k \times h = (3/7) \times 0.004 \times 10 = 0.017$ ft³ per min per foot of length of the cofferdam.

b. From the dimensions shown in Figure 3.18, the soil prism to be considered is 6 ft high by 3 ft wide in cross section. At the bottom of the prism, the head loss at the left (sheet pile) side is 4/7 of 10 ft and at the right side it is about 4.8/7 of 10 ft. The average remaining head is the head loss in the prism. This is equal to

$$i_{av} = 7 - (4 + 4.8)/2 = 7 - 4.4 = 2.6 \text{ ft}$$

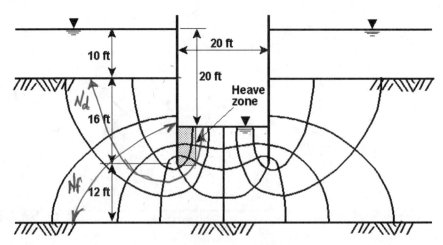

FIGURE 3.18 Flow net under cofferdam to accompany Problem 3.13.

The buoyed unit weight of the soil in the prism is

$$\gamma' = (120 - 62.4) \times 6 = 346 \text{ lb}$$

Thus, the factor of safety against heave, Equation 3.31, is

$$FS = \frac{\gamma'}{i_{av}\gamma_w} = \frac{346}{2.6 \times 62.4} = 2.1$$

REFERENCES

1. Hazen, A. 1892. "Physical properties of sands and gravels with reference to their use in filtration," *Report to the Massachusetts State Board of Health*, p. 539.
2. NAVFAC DM 7.01. 1982. "Soil Mechanics", Design Manual 7.01, Naval Facilities Engineering Command. Department of the Navy, Alexandria, VA. May.
3. Zangar, C. N. 1953. "Theory and Problems of Water Percolation, " Engineering Monograph No. 8. U.S. Department of Interior, Bureau of Reclamation. Denver, CO. April.
4. U.S. Army. 1983. "Dewatering and Groundwater Control, " Technical Manual TM 5-818- 5, NAVFAC P-418, AFM 88-5, Chap. 6. Departments of the Army, the Navy, and the Air Force. Washington, DC. November.
5. Peck, R. B. 1959. Class notes in "Foundation Engineering," a course presented by the Department of Civil Engineering, University of Illinois, Urbana-Champaign, IL.
6. Peck, R. B., Hanson, W. E., and Thornburn, T. H. 1974. *Foundation Engineering*, 2nd Edition, New York: John Wiley and Sons.
7. Dupuit, J. 1863. *Études théoriques et practiques sur le mouvement des eaux dans les canaux découveris et à travers les terrains perméables* (Theoretical and experimental studies of the flow of water in open channels and through permeable ground), 2nd ed. Paris: Dunod.
8. Casagrande, A. 1935. "Seepage through Dams, " *Journal of the New England Water Works Association,* Vol. 51, No. 2.
9. Terzaghi, K. 1943. *Theoretical Soil Mechanics*. New York: John Wiley and Sons. p.257.

CHAPTER 4
COMPRESSIBILITY OF SOIL AND ROCK

This chapter contains four topics related to the compressibility of soil and rock. First, we will look at the compressibility of a soil acted on by a pressure performing uniformly throughout its mass. Next, we will examine the distribution of pressure with depth in a soil mass acted on by a surface pressure of finite dimensions. These concepts then will allow us to evaluate the compression settlement of a foundation element or embankment. Finally, we will consider the time-rate at which the settlement will occur.

COMPRESSIBILITY

Consider a very wide, but relatively thin, layer of compressible soil between two porous but incompressible layers as shown in Figure 4.1.

Assume that the soil (a) is saturated, (b) is in effective stress equilibrium with the weight of the overlying layer, and (c) has never been subjected to a vertical pressure greater than that which now exists from the overlying layer. Then, assume that a uniform surcharge pressure is applied to the entire surface in increments, with each increment being allowed to come to effective stress equilibrium before the next load increment is applied. This creates a condition of one-dimensional compression. There is no horizontal deformation of the soil except near the boundaries of the loaded area.

This discussion is applicable mainly to transported and saturated silts and clays wherein the soil structure and composition are relatively uniform throughout the layer. Settlement, or volume change, due to one-dimensional compression results only from a decrease in the volume of voids (see phase diagrams in Chapter 2) and is analyzed in terms of effective stresses.

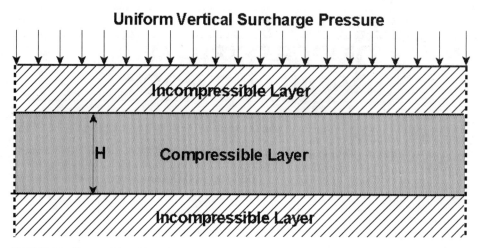

FIGURE 4.1 Compressible soil layer between two porous and incompressable layers, acted upon by a uniform vertical surcharge pressure.

One-Dimensional Compression

Figure 4.2 illustrates the one-dimensional compression effective stress versus the strain behavior of the soil layer, in terms of load and deformation.

In this example, it is assumed that the initial overburden pressure on the soil layer, before application of the surcharge, is 0.25 tons per square foot (24 kPa) and that the average void ratio in the soil layer is 1.30.

The behavior of the compressible soil layer is best understood if the soil is considered to be a *lead sponge*. At low overburden pressure, the soil initially has an open, high void ratio and a low strength structure of interconnected strands. As an increment of pressure is applied, the structure of the soil compresses plastically, forcing water out of the pore spaces, and the framework becomes stronger. Compression stops when the increasing structural strength is able to support the additional pressure. With continued pressure increases, the structural strength of the soil mass increases non-linearly at an ever-increasing rate in the form of a mathematical power curve, Figure 4.2. Because this load-deformation curve represents the behavior of a soil that has never been subjected to this intensity of loading before, it is referred to as the *virgin compression curve*.

Whenever the load is removed, the plastically deformed material (lead sponge) does not return to its original thickness, but instead rebounds only a small elastic amount, as shown in Figure 4.2. On reloading, the deformation of the material follows a hysteresis loop until it reaches the previously applied pressure, after which it resumes the initial curved relationship with pressure.

The curved relationship is cumbersome to analyze mathematically. Therefore, it is more common to represent the pressure-deformation curve of Figure 4.2 on a semi-logarithmic plot as shown in Figure 4.3, in which the effective stress is plotted on the horizontal axis, on a logarithmic scale, and the void ratio is plotted arithmetically on the vertical axis.

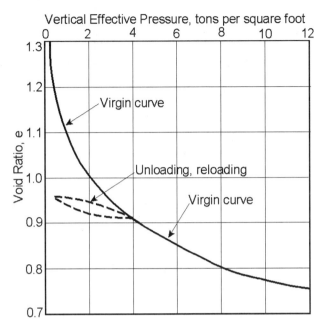

FIGURE 4.2 Arithmetic scale graph of effective stress vs. deformation curve for compressible soil, showing unloading-reloading cycle.

The virgin compression curve plots as a straight line whose slope is the *compression index*, C_c:

$$C_c = \frac{\Delta e}{\Delta \log p} = \frac{e_1 - e_2}{\log p_2 - \log p_1} \qquad (4.1)$$

where p_1 and p_2 = Effective stress points on the virgin compression curve
e_1 and e_2 = Void ratios corresponding to p_1 and p_2, respectively.

In the example of Figure 4.3, the soil initially was loaded in increments to 2.0 tons per square foot (190 kPa). Then, the surcharge load was removed in increments and the rebound, or swelling, curve was determined. Although it normally has a slight curvature, the swelling curve can be approximated by a straight line whose slope is the *swelling index* C_s, which is defined by an equation similar to Equation 4.1, except that the pressures are those at the beginning and end of the swelling cycle.

Next, the former surcharge load is reapplied to the soil in increments and further incremental loading is continued to a higher amount. The resulting deformation is plotted on Figure 4.3. The recompression curve roughly parallels the swelling curve up to the former maximum pressure, now known as the *preconsolidation pressure*. This is the maximum pressure to which the soil layer has ever been subjected. At surcharge pressures above the preconsolidation pressure, the void ratio versus the log pressure curve now resumes its original virgin compression slope and position on the graph.

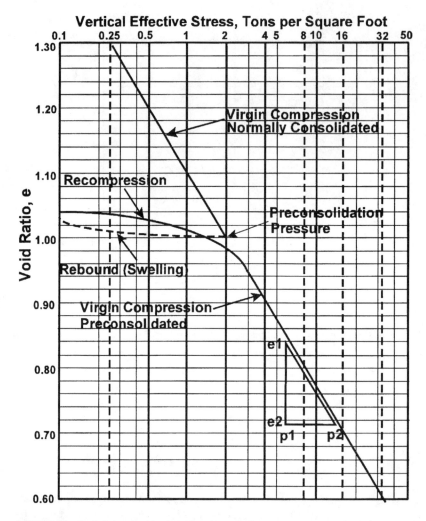

FIGURE 4.3 Typical plot of void ratio vs. logarithm of effective vertical pressure for compressible soil.

The curved slope of the recompression curve also can be approximated by a straight line. The slope of the line for vertical stress values below the preconsolidation pressure is the *recompression index* C_r, defined by an equation similar to Equation 4.1 and the equation for the swelling index. Values of the swelling and recompression indices are similar, differing only because of the hysteresis inherent in the process.

Laboratory Oedometer Test

In practice, the relationship between the void ratio and the logarithm of effective pressure is determined by testing an undisturbed sample of the soil in a laboratory oedometer, or consolidometer. The oedometer sample might have been part of a hand-carved or thin-wall tube sample, both of which are relatively undisturbed (see Chapter 7). A typical oedometer arrangement is shown in Figure 4.4.

The trimmed undisturbed sample, typically 2.5 in. (6.4 cm) in diameter by 1.0 in. (2.5 cm) thick, is contained in a metal ring with porous stones top and bottom to permit drainage. The oedometer is kept full of water to maintain sample saturation during the test period. Axial load is applied either by dead weights or controlled air pressure. Individual load increments are applied in a doubling ratio, such as a 0.25, 0.5, 1.0, 2.0, 4.0, 8.0, etc. tsf ratio, so that the results for the virgin compression curve will plot with evenly spaced points on a semi-logarithmic plot. Individual load increments are typically maintained for at least 24 hours each to permit the complete expulsion of pore water and the development of fully effective stress on the soil structure.

At least once during the loading process, typically after virgin compression has been reached, an unload-reload sequence is used to define the swelling index. The sample is unloaded incrementally to zero, or to a small pressure, and then reloaded in increments, with each increment held until swelling or recompression ceases. This permits a laboratory determination of the swelling index and the recompression index.

Determining Preconsolidation Pressure

Invariably, the soil layer from which a sample is taken has experienced some amount of effective overburden pressure above the existing pressure during its geological history. This might have been from a layer of soil that was later removed, either by natural erosion or by man, or from desiccation. During sampling, the soil sample swells slightly because of the release of the existing overburden pressure. Then, during the laboratory consolidation test, a small amount of recompression will occur when the applied effective pressure

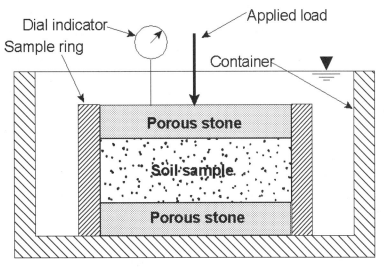

FIGURE 4.4 Laboratory oedometer, or consolidometer, for determining void ratio vs. effective pressure relationship for soil samples.

is less than the preconsolidation pressure. When the applied effective pressure becomes greater than the preconsolidation pressure, the logarithm of effective pressure curve (e-log p curve) becomes steeper and represents virgin compression of the sample.

Three terms related to cohesive soil (clays and silty clays) now can be defined:

1. *Normally consolidated soil* has a current effective overburden pressure equal to the preconsolidation pressure.
2. *Overconsolidated soil* has a current effective overburden pressure less than the preconsolidation pressure.
3. The *overconsolidation ratio*, p_c / p_o, is the ratio of the preconsolidation pressure to the current effective overburden pressure.

The preconsolidation pressure can be determined by a graphical procedure first suggested by Casagrande in 1936 [1].

The procedure is illustrated in Figure 4.5 and progresses in the following manner:

1. By visual observation, draw line c-d tangent to the e-log p curve at the point of maximum curvature, touching at point a. The slope of the line should be such that the area between line c-d and the e-log p curve is the same on both sides of point a.

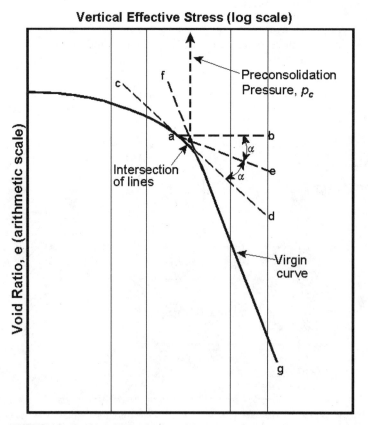

FIGURE 4.5 Graphic procedure for determining preconsolidation pressure.

2. Draw a horizontal line *a-b*.

3. Draw the line *a-e* to bisect the angle between lines *a-b* and *a-d*; that is, the two angles α should be equal.

4. Project the straight line of virgin compression from *g* to *f*. The intersection point of lines *a-e* and *f-g* is at the preconsolidation pressure.

Determining the Field Void Ratio-Pressure Relationship

Field e-log p Curve. A relatively undisturbed sample very likely has experienced some disturbance from the sampling, transport, and handling process, resulting in a laboratory virgin compression index, shown by the slope of the virgin curve, that is somewhat flatter than it would be if a truly undisturbed sample were tested.

The procedure shown in Figure 4.6 was developed by J. Schmertmann in 1953 [2]. He found from laboratory tests that the virgin compression curves for a series of samples at varying degrees of remolding converge at 0.42 times the initial void ratio. He also concluded from empirical evidence that the initial recompression curve, line *a-b* of Figure 4.6,

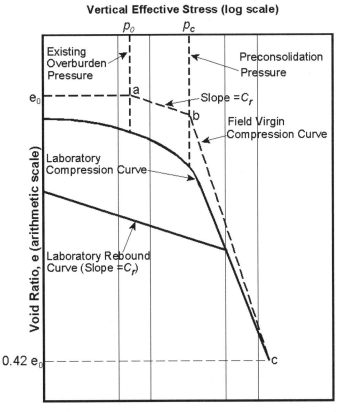

FIGURE 4.6 Determination of probable field void ratio-effective pressure curve from laboratory sample of low-to-medium sensitivity compressible soil.

is parallel to the rebound curve from the laboratory oedometer test. Therefore, we can establish the field consolidation curve for truly undisturbed soil from the laboratory curve as follows:

1. Compute the initial void ratio e_0 for the sample, using the equations in Chapter 2.
2. Estimate the effective overburden pressure p_0 now existing on the sample. Draw a horizontal line from e_0 on the vertical axis to p_0 at point a.
3. Draw the sloping line parallel to the laboratory rebound curve from a to b. The point b is at the vertical line of the preconsolidation pressure.
4. Extend the laboratory virgin compression curve to a point $0.42e_0$ and then draw the field virgin compression curve from point b to the intersection point.

Estimating the Compression Index. Often, laboratory consolidation tests are not readily available because of their high cost. In those instances, a reasonable estimate of the compression index can be obtained from the published research literature. Some published correlations for the compression index C_c adapted from several sources are contained in Table 4.1.

TABLE 4.1. Some published correlations for Compression Index, C_c* (adapted from several published sources).

Equation	Regions of applicability
$C_c = 0.007 \, (LL - 10) \pm 30\%$	Remolded clays of low to medium sensitivity
$C_c = 0.009 \, (LL - 10) \pm 30\%$	Undisturbed clays of low to medium sensitivity
$C_c = 0.5 \, (\text{Plast. Index}) \times (\text{Specific Gravity of Solids})$	Remolded clays of low to medium sensitivity
$C_c = 0.208 \, e_0 + 0.0083$	Chicago clays
$C_c = 17.66 \, (10^{-5} w_n^2 + 5.93 \, (10^{-3}) w_n - 0.0135$	Chicago clays
$C_c = 0.01 \, w_n$	Chicago clays
$C_c = 1.15 \, (e_0 - 0.27)$	All clays
$C_c = 0.0115 \, w_n$	Organic soils-meadow mats, peats, and organic silt and clay
$C_c = 0.30 \, (e_0 - 0.27)$	Inorganic, cohesive soil; silt with some clay; silty clay; and clay
$C_c = 0.75 \, (e_0 - 0.50)$	Soils of very low plasticity
$C_c = 0.156 \, e_0 + 0.0107$	All clays
$C_c = 0.0046 \, (LL - 9)$	Brazilian clays

*e_0 = in situ void ratio; w_n = in situ water content

Estimating the Swelling Index. The swelling index C_s is much smaller than the compression index and is determined from the laboratory oedometer test. In most cases, however, $C_s \approx 0.2$ to $0.1\ C_c$. If the liquid limit is known, Figure 4.7 can be used to estimate the swelling index by using the void ratio from which the rebound occurs.

Estimating the Preconsolidation Pressure. Several correlation methods exist to estimate the preconsolidation pressure. If the Atterberg limits (liquid and plastic limits) and an estimate for the unconfined compressive strength (see Chapter 5) are available, then a reasonable estimate of the preconsolidation pressure can be made from

$$P_c = \frac{0.5 q_u}{0.11 + 0.0037\ PI} \quad (4.2)$$

in which q_u is the unconfined compressive strength and *PI* is the plasticity index.

Calculation of Settlement Due to One-Dimensional Primary Consolidation

Based on the information obtained from a laboratory oedometer test of a relatively undisturbed sample, we can now calculate the probable settlement due to one-dimensional primary consolidation under a uniform surcharge pressure.

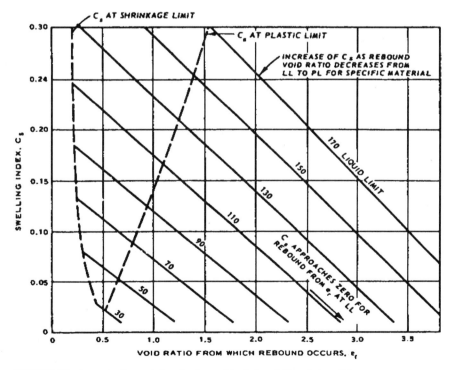

FIGURE 4.7 Approximate correlations for the swelling index of silts and clays. [3]

Consider a saturated clay layer, such as that shown in Figure 4.1, which has a thickness H and is at equilibrium under an effective pressure p_1 due to the overlying layer. The change in thickness, ΔH, due to the addition of an effective overburden pressure p_2, is the settlement S. The change in thickness for a given area is equal to the change in volume of voids, ΔV_v, which in turn is equal to $\Delta e \cdot V_s$ where Δe = change of void ratio. From this we can, by using the equations given in Figure 2.5, Chapter 2, arrive at:

$$S = H \frac{\Delta e}{1 + e_o} \quad (4.3)$$

Then, for the virgin compression portion of the curve of Figure 4.3 we combine Equations 4.1 and 4.2, as follows:

$$S = \frac{HC_c}{1 + e_o} \log \left(\frac{p_2}{p_1} \right) \quad (4.4)$$

If the clay layer is thicker, or if it contains several sub-layers with different consolidation characteristics, the calculations can be made for each sub-layer and summed as:

$$S = \sum \left[\frac{H_i C_{c(i)}}{1 + e_{o(i)}} \log \left(\frac{p_{2(i)}}{p_{1(i)}} \right) \right] \quad (4.5)$$

where H_i = Thickness of sub-layer i
$C_{c(i)}$ = Compression index of sub-layer i
$p_{1(i)}$ = Initial average effective overburden pressure on sub-layer i
$p_{2(i)}$ = Final average effective overburden pressure on sub-layer i.

For an overconsolidated soil, such as that illustrated in Figure 4.6, Equations 4.3 and 4.4 can be applied to the recompression curve as well as to the virgin curve, simply by substituting C_s for C_c at the appropriate locations. This will be illustrated in Problem 4.1 below.

VERTICAL STRESSES DUE TO SURFACE LOADS

Boussinesq used the theory of elasticity to solve the theoretical problem of stresses produced at any point in the interior of a homogenous, isotropic, elastic medium by a point load on the surface of an infinitely large space [4]. Westergaard presented similar equations for thin, laterally reinforced layers [5].

Boussinesq Equations

Starting with Boussinesq's development, others have extended the work to uniformly loaded lines and areas such as infinitely long strips, circles, and rectangles. The solutions contain vertical pressure, horizontal pressure, and shear stress at a point at a given depth and a given horizontal distance from the origin. Because stress increases are in the elastic region, stresses from several sources can be superimposed on each other.

Solutions have been developed for a number of practical situations, several of which are presented below. Terzaghi [6] empirically modified the Boussinesq equations for estimating the horizontal pressures on rigid retaining structures (see Chapter 10) due to point loads, line loads, and strip loads.

Point Load. A wheel, or any concentrated load on a small area, can be treated as a point load. The lateral pressure varies with depth and horizontal distance from the load. The pressure is greatest along a vertical line closest to the load. Figure 4.8 presents the empirical Terzaghi equations and accompanying loading diagram [6].

Uniformly Loaded Line. A continuous wall footing of narrow width or a loaded rail line can be taken as a line load when located parallel to the retaining structure. The lateral pressure increases from zero at the ground surface to a maximum at a certain depth and then gradually diminishes to zero at a greater depth. Figure 4.9 presents the empirical Terzaghi equations and accompanying loading diagram [6].

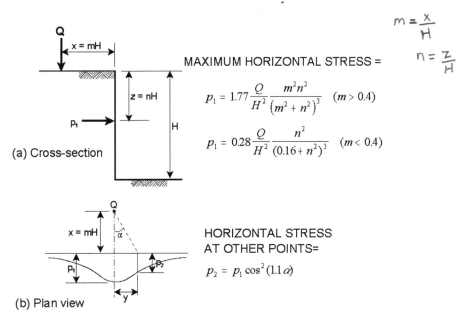

FIGURE 4.8 Point load on rigid retaining structure using Boussinesq equations empirically modified by Terzaghi, [6]. (Adapted from Teng, [7])

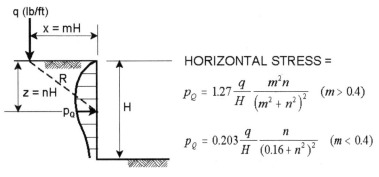

FIGURE 4.9 Line load on rigid retaining structure using Boussinesq equations, empirically modified by Terzaghi, [6]. (Adapted from Teng, [7]).

Uniformly Loaded Strip. A highway, railroad, or continuous wall footing parallel to the retaining structure can be treated as a strip load. Similar to that of the line load, its lateral pressure increases from zero at the ground surface to a maximum at a certain depth and then gradually diminishes to zero at a greater depth. Figure 4.10 presents the empirical Terzaghi equations and accompanying loading diagram [6].

Theoretical solutions of the Boussinesq equations have been summarized in charts, such as the one shown in Figure 4.11, that give reasonable approximations in real soils for such uniformly loaded structures as liquid storage tanks and embankment fills. Charts to facilitate calculations for additional conditions, including those below, are available in the geotechnical engineering literature.

Uniformly Loaded Rectangular Area. Loads applied to a soil normally are distributed over an area. Small loaded areas often are located so closely that vertical stresses overlap. Because of assumed elasticity, calculated pressures can be superimposed upon each other to reach a solution. The Boussinesq solution for the vertical pressure under the corner of a uniformly loaded rectangular area is shown in Figure 4.12.

By using adding areas to make a full rectangle and then subtracting the zero-load areas, a vertical pressure below any point can be found. This will be illustrated in Problem 4.2 below.

Uniformly Loaded Circular Area. For a uniformly loaded circular area, the Boussinesq solution for the vertical pressure under any point under or outside the area is shown in Figure 4.13. This will be illustrated in Problem 4.3 below.

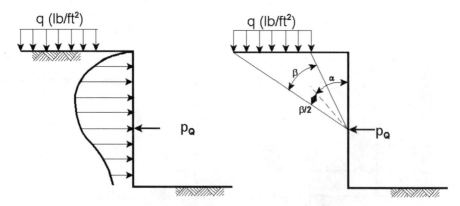

HORIZONTAL STRESS AGAINST RIGID WALL=

$$p_Q = \frac{2q}{\pi}\left[(\beta + \sin\beta)\sin^2\alpha + (\beta - \sin\alpha)\cos^2\alpha\right]$$

FIGURE 4.10 Strip load on rigid retaining structure using Boussinesq equations, empirically modified by Terzaghi, [6]. (Adapted from Teng, [7]).

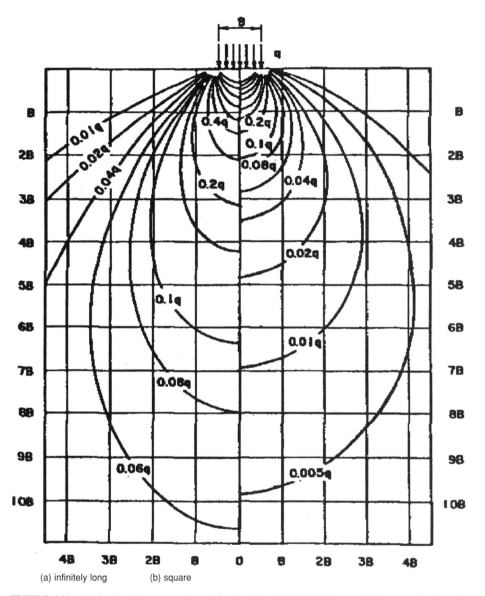

FIGURE 4.11 Disribution of pressure under uniformly loaded (a) infinitely long and (b) square footings using Boussinesq distribution [8].

Uniformly Loaded Embankment. Embankment loads made of compacted fill are commonly used. Such loads satisfy the requirement for a flexible load and the Boussinesq solution for the vertical pressure under any point under or outside the area is shown in Figure 4.14.

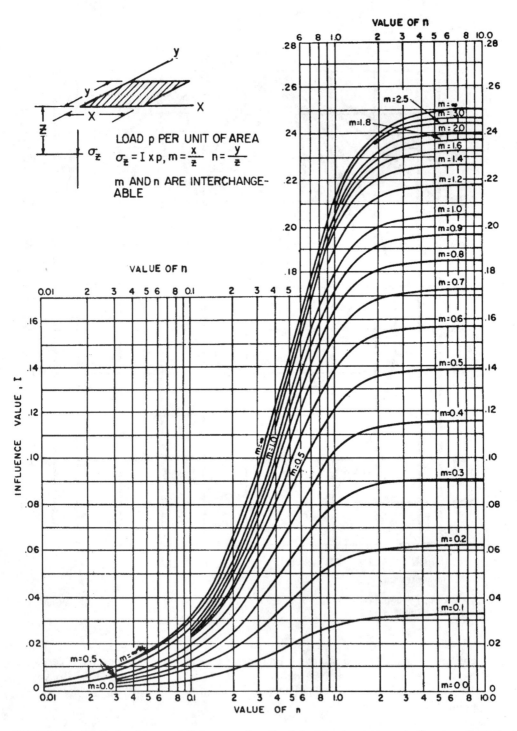

FIGURE 4.12 Distribution of pressure under the corner of a uniformly loaded rectangular area using Boussinesq distribution [3].

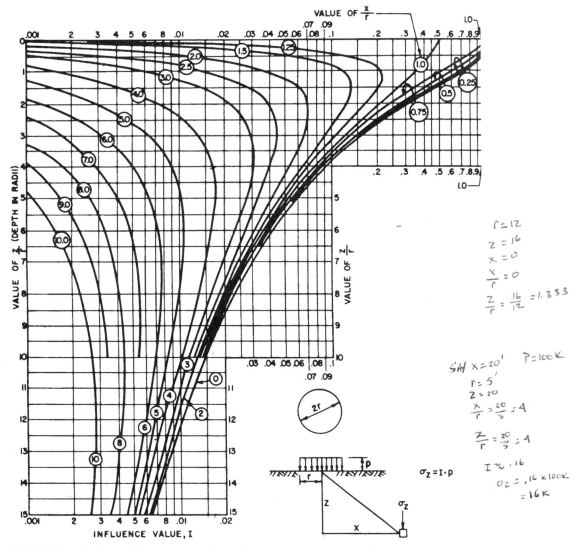

FIGURE 4.13 Distribution of pressure under uniformly loaded circular area using Boussinesq distribution [8].

Stresses Under Rigid Footings

The real pressure distribution under a rigid foundation, as opposed to that of a flexible foundation, is shown schematically in Figure 4.15.

Real soils are not elastic, not homogeneous, and not of infinite depth. This disparity indicates why the Boussinesq equations do not indicate the real pressure distribution under some of the more common foundation elements.

The *2:1 Rule* shown in Figure 4.16 is commonly used as a practical approximation.

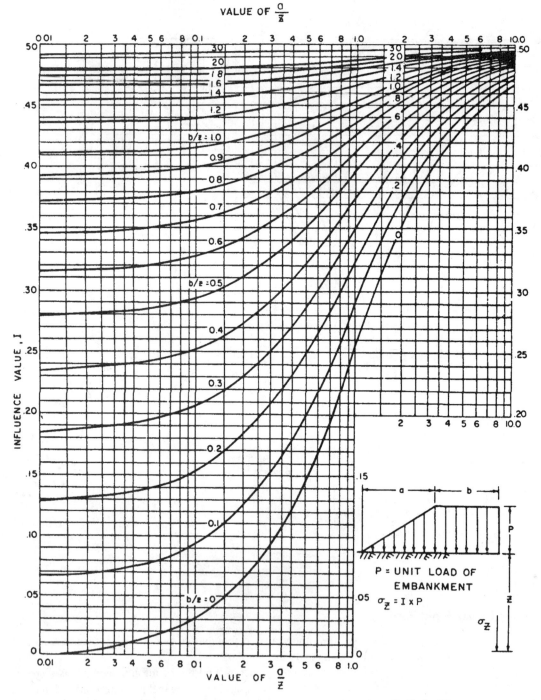

FIGURE 4.14 Distribution of pressure under uniformly loaded embankment using Boussinesq distribution [8].

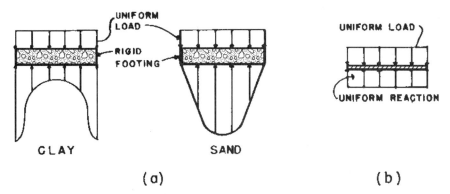

FIGURE 4.15 Contact pressure under (a) rigid footings and (b) flexible foundation on an elastic half space [8].

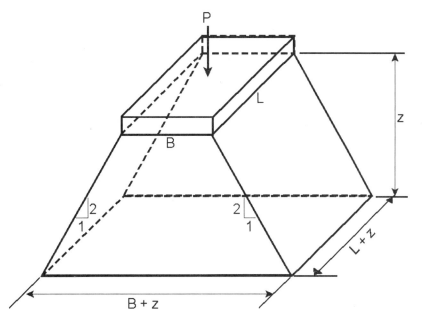

FIGURE 4.16 The 2:1 Rule for approximating the vertical stress distribution under a rigid foundation is on top of a truncated pyramid of base $B + z$ and $L + z$.

This rule assumes that the applied surface load spreads downward uniformly at an angle of about 63.4° from the horizontal, corresponding to a slope of 2 vertical to 1 horizontal, parallel to both sides. At any depth z below the footing, the rectangle is increased by z in width on each side, and the surface force P is distributed over an area $(B + z) \times (L + z)$.

CONSOLIDATION SETTLEMENT OF STRUCTURES

Foundation Settlement in Cohesive Soil

In practice, rigid foundations of limited dimensions are often founded in compressible, cohesive (clayey) soils having several layers. In such cases, it is common to

1. Determine the location, thickness, and compressibility properties of each layer, i.e., e_o, p_1, C_c, and p_c, either by obtaining undisturbed samples and making laboratory oedometer tests and/or by using correlations with index properties to estimate the pertinent properties.
2. Use the concept of Equation 4.5; that is, subdivide the substrata into a number of sublayers, based on natural stratification and on the preconsolidation load within each layer.
3. Determine the pressure increase profile under the footing using the 2:1 Rule.
4. Determine the average increase in vertical pressure on each layer by

$$\Delta p = \Delta p_{avg} = \frac{\Delta p_t + 4\Delta p_m + \Delta p_b}{6} \qquad (4.6)$$

where Δp_t, Δp_m, and Δp_b are the increases in pressure at the top, middle, and bottom of the layer, respectively.

5. Calculate the void ratio change for each sublayer, and then calculate the increment of settlement due to that sublayer using Equation 4.3. Alternately, use the compression index C_c, and calculate with Equation 4.4. Sum up the incremental settlements for all layers.

This procedure is illustrated in Problems 4.4 and 4.5.

Foundation Settlement in Cohesionless Soil

The settlement of shallow foundations in cohesionless soil, which is essentially sand, follows the same theoretical basics as do foundations on cohesive soil, except that (a) the compression index is much flatter, (b) the preconsolidation pressure is a function of relative density, and (c) the time for complete consolidation (see below) is very short and typically occurr during construction. This topic will be discussed at greater length in Chapter 8.

TIME RATE OF CONSOLIDATION

Theoretical Development

Considering the concept of a lead sponge presented above and the usual spring analogy presented in most soil mechanics textbooks, it is apparent that a saturated soil does not reach effective stress equilibrium immediately upon the application of an increment of load. The rate at which a drop of water will move to the nearest exit point depends on the permeability of the soil and on the hydraulic gradient. The permeability of a clean sand is many thousands times that of a clay.

When the load is first applied, the pressure in the pore water causing the hydraulic gradient is high; at the instant of load application it is equal to the applied external pressure. Then, as pore water is squeezed out, the soil skeleton gradually assumes increasingly more of the load and the hydraulic gradient forcing more water out becomes increasingly smaller. Furthermore, the decrease in pore-water pressure, or the degree of consolidation, is not uniform throughout the layer but rather follows a parabolic pattern with 100 percent consolidation, or full effective stress and zero pore-water pressure, at the drainage surface(s), and the full remaining pore-water pressure at the greatest distance from a drainage surface.

As developed by Terzaghi in 1925 [9], the permeability of a cohesive soil is related to

$$c_v = \frac{k(1+e)}{a_v \gamma_w} \quad (4.7)$$

where c_v = Coefficient of volume compressibility, a property of the soil
       k = Coefficient of permeability
       a_v = Coefficient of compressibility
       γ_w = Unit weight of water.

The coefficient of compressibility a_v is the slope of the void ratio versus the effective pressure curve when plotted arithmetically as in Figure 4.2. The slope is taken as the average in a pressure increment between two successive pressures. Because the normal method of plotting compressibility data is the void ratio versus log pressure curve of Figure 4.3, a_v can be found from

$$a_v = \frac{0.435 \, C_c}{p} \quad (4.8)$$

where $p = (p_1 + p_2)/2$, the average pressure for the increment.

For a given set of boundary conditions the relationship for percent consolidation is

$$U_z (\%) = f\left(T_v, \frac{z}{H}\right) \quad (4.9)$$

in which

$$T_v = \frac{c_v}{H^2} t \quad (4.10)$$

where T_v = Time factor, a dimensionless number
       H = Half-thickness of the soil layer, or the distance to the nearest drainage face
       t = Time corresponding to the degree of consolidation U_z.

The symbolic relationship of Equation 4.9 is tabulated in Table 4.2, and pictured in Figure 4.17 for three different conditions of the distribution of pore-water pressure with depth. Case 1, typically found in foundation engineering practice, is for the condition in which the consolidation pressure is constant with depth, as in a thin layer, or decreases with depth, as in a thick layer. Case 2 represents the condition of a sinusoidal pressure distribution. Case 3 is indicative of a hydraulic fill in which the only pressure is its own self-weight

In the first case, the consolidation pressure is constant with depth, as in a thin layer, or decreases with depth, as in a thick layer. The second case represents the a sinusoidal pressure distribution. The third case is indicative of a hydraulic fill in which the only pressure is its own weight. The first case is the condition typically found in foundation engineering practice.

TABLE 4.2 Degrees of Consolidation as a Function of Time Factor Tv

T_v	Average Degree of Consolidation, U, percent		
	Case 1	Case 2	Case 3
0.004	7.14	0.98	0.80
0.008	10.09	1.95	1.60
0.012	12.36	2.92	2.40
0.020	15.96	4.81	4.00
0.028	18.88	6.67	5.60
0.036	21.40	8.50	7.20
0.048	24.72	11.17	9.69
0.060	27.64	13.76	11.99
0.072	30.28	16.28	14.36
0.083	32.51	18.52	16.51
0.100	35.68	21.87	19.77
0.125	39.89	26.54	24.42
0.150	43.70	30.93	28.86
0.175	47.18	35.07	33.06
0.200	50.41	38.95	37.04
0.250	56.22	46.03	44.32
0.300	61.32	52.30	50.78
0.350	65.82	57.83	56.49
0.400	69.79	62.73	61.54
0.500	76.40	70.88	69.95
0.600	81.56	77.25	76.52
0.800	88.74	86.11	85.66
1.000	93.13	91.52	91.25
1.500	98.00	97.53	97.45
2.000	99.42	99.28	99.26

Laboratory Consolidation Test

Primary Consolidation. The determination of the soil consolidation parameters, Equations 4.7 through 4.10, for determining the time rate of consolidation of a clay layer, is normally done with a laboratory consolidation test on an undisturbed sample from the layer. A typical laboratory oedometer (consolidometer) is shown in Figure 4.4. The test usually is conducted in the following manner:

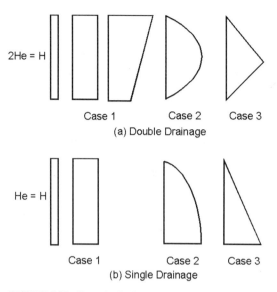

FIGURE 4.17 Example distribution of excess pore water pressure for double and single drainage. H is actual layer thickness and H_e is the equivalent thickness.

1. An initial, very small static load is applied to the sample to seat the components and the dial indicator is set to zero (or a convenient starting number).

2. The first load increment, for example, 0.25 tons/ft², is applied and deformation dial readings are made at doubled increasing times, typically at 15 sec, 30 sec, 1 min, 2 min, 4 min, 8 min, 15 min, 30 min, 1 hr, 2 hr, 4 hr, 8 hr, and 24 hr (or longer if needed to define the secondary consolidation curve). This time sequence provides evenly spaced points on a deformation versus log time graph and also permits one increment of loading to be done in a technician work day of eight hours.

3. Successive increments of load are applied, with Step 2 repeated at each increment. The load sequence increases on a doubling basis to permit even spacing of points on the e-log p curve. A typical sequence, in tons per square foot, is 0.25, 0.5, 1.0, 2.0, 4.0, 8.0, 16.0, 32.0, and so forth until several points define the virgin compression curve.

4. Using each of the deformation versus log time plots for load increments above the preconsolidation pressure, determine the deformation dial reading at zero percent and 100 percent primary consolidation. Keep in mind that deformation versus log time plots are not well defined below the preconsolidation pressure. From these, determine the time corresponding to 50 percent primary consolidation (see method below). Solve for the coefficient of volume compressibility c_v using Equation 4.10, recognizing that the laboratory test is double drainage and, therefore, the value of H is one-half the height of the laboratory sample and, that at 50 percent consolidation, $T_v = 0.197$.

5. For samples that consolidate too rapidly to permit a timely development of the early part of the deformation versus log time curve, the square root of time fitting method shown in Figure 4.19 may be used; for 90 percent consolidation, $T_v = 0.848$.

6. Given the increment values of c_v and a_v (Equation 4.8) from the laboratory consolidation test, the coefficient of permeability k can be calculated for each increment using Equation 4.7.

The time rate of settlement for a vertical pressure on the surface of a clay layer now can be estimated using an average value of c_v for the corresponding pressure range in the laboratory test.

Log of Time Fitting Method. As shown in Figure 4.18, the early part of the time curve occurs very rapidly. Assuming this section of the curve is parabolic, the amount of deformation between two points—the distance between which is four times the time value—is doubled and the *zero* time deformation dial reading is set at two times the difference. In Figure 4.18, the deformation between *time* = 0.25 min and *time* = 1.0 min is 0.0043 in. This is doubled to 0.0086 in and is measured up from the 1 min reading, for a dial reading of 159.

The straight-line center part of the curve is extended downward, as shown in Figure 4.18. The secondary consolidation portion of the curve is fitted with a straight line and that line is extended backward to cross the extended primary line. The cross point represents the deformation dial corresponding to 100 percent primary consolidation. Then, the difference in deformation dial readings for zero percent and 100 percent consolidation is halved and added to the zero reading to give the 50 percent consolidation point. The time corresponding to 50 percent t_{50} then is used in Equation 4.10.

$$c_v = \frac{0.197 H^2}{t_{50}} \tag{4.11}$$

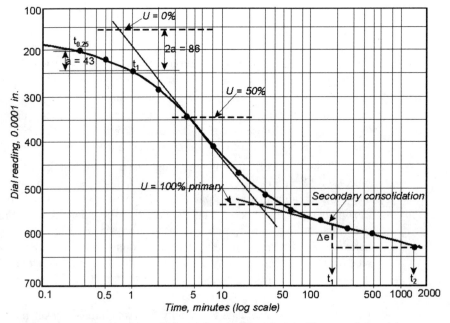

FIGURE 4.18 Typical deformation dial vs. log time curve for a single load increment during a laboratory consolidation test showing log time-fitting method and secondary consolidation.

Square Root of Time Fitting Method When appropriate, the dial reading versus time values for a load increment can be plotted as deformation versus the square root of time for the first few increments, as shown in Figure 4.19, in which the same deformation versus time values as those in Figure 4.18 were used.

A straight line is fitted to the first several points and that line is extended backward to zero time. Then, a second line having a slope 1.15 times the first line is drawn through zero. The point at which the second line crosses the deformation versus the square root of time curve is the 90 percent consolidation time t_{90}, which is used in Equation 4.10,

$$c_v = \frac{0.848 \, H^2}{t_{90}} \qquad (4.12)$$

Secondary Consolidation. The concept of 100 percent primary consolidation implies that the soil is now in equilibrium with the vertical pressure and the excess pore-water pressure is zero throughout the sample. Continued deformation after that time is called *secondary consolidation*, and is considered to result from a reorientation of soil grains causing further volume change at constant effective pressure. Secondary consolidation is shown on Figure 4.18. Using the deformation versus log time plot for each load increment, the *secondary compression index* can be expressed as

$$c_\alpha = \frac{\Delta e}{\log t_2 - \log t_1} \qquad (4.13)$$

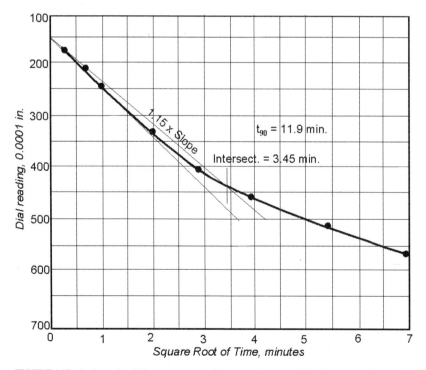

FIGURE 4.19 Deformation dial vs. square root of time curve for a single load increment during a laboratory consolidation test showing square root of the time-fitting method.

where t_1 and t_2 are times taken within the secondary consolidation section of the deformation versus log time curve of Figure 4.18 and Δe is the void ratio change corresponding to these two times. The magnitude of secondary consolidation can be calculated as

$$S_s = \frac{C_\alpha H}{1+ e_p} \log (t_2/t_1) \qquad (4.14)$$

Secondary consolidation is an important factor for organic and highly compressible soils. In most overconsolidated inorganic clays, the secondary compression index is very small and may be ignored.

EXAMPLE PROBLEMS

One-Dimensional Compression

4.1 Question:
Assume that the compressible soil shown in Figure 4.1 consists of a clay layer 10 ft thick. A laboratory oedometer test made on a sample from the center of the layer indicates the following:

Initial void ratio, $I = 1.10$,
Saturated density, $\gamma_{sat} = 115$ lb/ft^3
Existing overburden pressure $p_o = 0.7$ tons/ft^2
Preconsolidation pressure, $p_c = 2.2$ tons/ft^2
Laboratory compression index $C_c = 0.35$
Swelling index $C_s = 0.06$.

Calculate the settlement of the surface of the upper incompressible layer due to the imposition of a uniform surcharge pressure of 5.0 tons/ft^2.

Answer:
Note: all calculated values are shown on Figure 4.20.

a. First, calculate the void ratio change from present overburden p_1 to the preconsolidation pressure p_c. Transposing Equation 4.1, $e_2 = e_1 - C_s \log (p_2/p_1)$ and using the swell index, the void ratio e_2, at point $b = 1.10 - 0.06 \log (2.2/0.7) = 1.07$. This also gives the starting point for the field virgin compression curve.

b. Next, calculate the field virgin compression index, line *b-c*, using Equation 4.1

$$C_c = \frac{1.07 - 0.42}{\log (90/2.2)} = 0.40$$

c. Then find the void ratio change from the preconsolidation pressure to the surcharge pressure. Again, transposing Equation 4.1, $e_2 = e_1 - C_s \log (p_2/p_1)$ and using the derived field compression index, the final void ratio e_f, at point $d = 1.07 - 0.40 \log (5.0/2.2) = 0.93$.

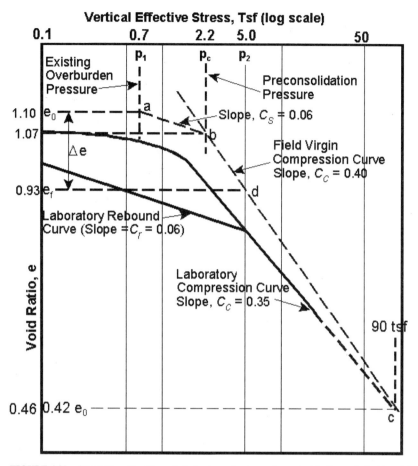

FIGURE 4.20 Calculation of settlement of clay layer due to surcharge load in Example Problem 4.1.

d. Calculate the settlement using Equation 4.3. With a surcharge load of up to 2.2 Tsf (the preconsolidation pressure), the void ratio change $\Delta e = 1.10 - 1.07 = 0.03$. The total settlement caused by the full surcharge load is $\Delta e = 1.10 - 0.93 = 0.17$.

$$S = H \frac{\Delta e}{1 + e_o} = 10 \times 12 \times \frac{0.17}{1 + 1.10} = 9.7 \text{ in.}$$

Vertical Stresses Due to Surface Loads

4.2 Question:
Assume that the rectangular mat footing shown in Figure 4.21 has dimensions of 4.0 m by 8.0 m. Calculate the vertical stresses at a depth of 4.0 m along a diagonal line from the center to a distance of 1.0 m in each direction outside the corner due to a uniform vertical footing pressure of 100 kPa. Use the Boussinesq solution in Figure 4.12 to plot the pressure

distribution. Then, use the 2:1 Rule of Figure 4.16 to estimate the pressure under the foundation mat and plot your results on Figure 4.21.

Answer:
For the four cases shown in Figure 4.21, use superposition to solve, as follows:

(a) For rectangle *AGOE*, $m = 2/4 = 0.5$; $n = 4/4 = 1.0$. From Figure 4.12, $I\sigma = 0.120$. By superposition, add the stresses from the four quarter sections and find that the stress at O is $4(100) 0.120 = 48.0$ kPa

(b) Use four different-sized rectangles meeting at O and add by superposition.

For rectangle *AGOE*, $m = 3/4 = 0.75$; $n = 6/4 = 1.5$; $I\sigma = 0.170$.
For rectangle *GBFO*, $m = 3/4 = 0.75$; $n = 2/4 = 0.5$; $I\sigma = 0.107$.
For rectangle *FCHO*, $m = 1/4 = 0.25$; $n = 2/4 = 0.5$; $I\sigma = 0.047$.
For rectangle *HDEO*, $m = 1/4 = 0.25$; $n = 6/4 = 1.5$; $I\sigma = 0.073$.
Stress at $O = 100 (0.170 + 0.107 + 0.047 + 0.073) = 39.7$ kPa.

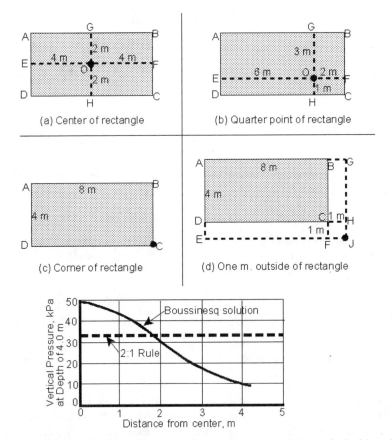

FIGURE 4.21 Distribution of pressure under rectangular mat foundation at a depth of 4 m using Boussinesq case and the 2:1 Rule in Example Problem 4.2.

(c) For rectangle *DABC*, $m = 8/4 = 2.0$; $n = 4/4 = 1.0$; $I\sigma = 0.200$.

  Stress at $C = 100 (0.200) = 20.0$ kPa

(d) To find the stress at a point outside the specified rectangle, add enough additional rectangles to reach the point as a corner. Then, subtract the effect due to the added rectangles. Note that one small area is subtracted twice, so it must be added back.

  For rectangle *EAGJ*, $m = 9/4 = 2.25$; $n = 5/4 = 1.25$; $I\sigma = 0.216$.
  For rectangle *EDHJ*, $m = 9/4 = 2.25$; $n = 1/4 = 0.25$; $I\sigma = 0.076$.
  For rectangle *FBGJ*, $m = 1/4 = 0.25$; $n = 5/4 = 1.25$; $I\sigma = 0.071$.
  For rectangle *FCHJ*, $m = 1/4 = 0.25$; $n = 1/4 = 1.5$; $I\sigma = 0.027$.
  Stress at $O = 100 (0.216 - 0.076 - 0.071 + 0.027) = 9.6$ kPa.

Plotted values for stress distribution along the diagonal are shown on Figure 4.21

4.3 Question:

A circular oil tank is 50 ft in diameter. The uniform pressure on the bottom is 500 lbs/ft^2. For a vertical line through the center and one through an edge, draw the distribution of pressure to a depth of 100 ft. Then, use the 2:1 Rule of Figure 4.16 to estimate the pressure under the oil tank.

Answer:

The solution is found using the Boussinesq solution in Figure 4.13 and the 2:1 Rule. The tank is on top of a truncated cone. The results are plotted on Figure 4.22.

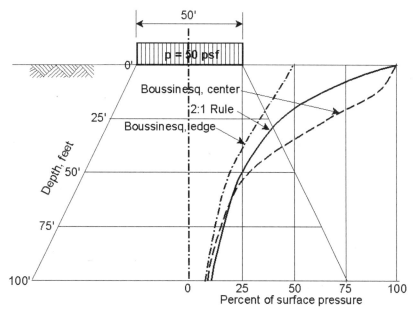

FIGURE 4.22 Distribution of vertical pressure under the oil tank using Boussinesq solution and the 2:1 Rule to accompany Example Problem 4.3.

Foundation Settlement in Cohesive Soil

4.4 Question:

Calculate the settlement of the shallow footing shown in Figure 4.23.

The water table is at the surface of the lower, clay layer. The compressibility characteristics of the two soil layers were determined from test boring and laboratory information from disturbed samples and from a reasonable estimate of *in situ* density: Assume the soils are normally consolidated.

Answer:

Follow the procedure given above under *Foundation Settlement in Cohesive Soil*. Summarize all values in Table 4.3.

(a) Estimate the unit weight and moisture content of each soil layer and estimate the initial void ratio. Separate the soil strata into layers for better accuracy.

(b) Calculate the existing (in situ) effective stress at the mid-point of each sublayer.

(c) Using the 2:1 Rule, calculate the increased pressure at the top, middle, and bottom of each sub-layer due to the net footing pressure (ignore the difference in unit weight between that of the soil and the concrete).

(d) Determine or estimate the Atterberg limits. Using the equations of Table 4.1, estimate the compression index C_c.

(e) Calculate the settlement of each sublayer using Equation 4.4 and get the total using Equation 4.5.

(f) In this example, $S = 1.80 + 1.61 + 0.91 + 0.49 = 4.7$ in.

4.5 Question:

Calculate the settlement of the shallow footing of Problem 4.4, assuming the soils are preconsolidated. This is determined from the unconfined compressive strength of each layer using Equation 4.2. The unconfined compressive strength is either estimated by field tests (see Chapter 6) or determined from laboratory tests of undisturbed tube samples.

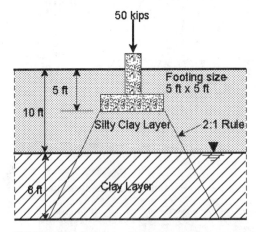

FIGURE 4.23 Settlement of footing on clay, to accompany Example Problem 4.4.

TABLE 4.3 Calculations to Accompany Example Problems 4.4 and 4.5.

Property	Silty Clay, 5 ft to 7 ft	Silty Clay, 7 ft to 10 ft	Clay, 10 ft to 14 ft	Clay, 14 ft to 18 ft
Layer thickness, H, inches	24	36	48	48
Unit weight, lbs./cu. ft.	128	128	110	115
Initial void ratio, e_0	0.64	0.64	1.00	0.90
Liquid limit	40	40	52	52
Plasticity index	19	19	30	30
Initial vertical pressure, p_1, mid-depth of layer. lbs./sq. ft.	6 × 128 = 768	8.5 × 128 = 1088	2(110-62.4) + 10 ×128 = 1375	4(110-62.4) + 2(115-62.4) + 1280 = 1576
Net footing contact pressure, lbs./sq. ft.	50000 / 25 = 2000 psf			
Increased pressure, top of layer, p_t	2000	1020	500	255
Increased pressure, center of layer, p_m	1389	889	347	195
Increased pressure, bottom of layer, p_b	1020	500	255	154
Average increased pressure, Δp, psf	1429	846	357	198
Total footing pressure, $p_2 = p_1 + \Delta p_{avg}$	2197	2034	1732	1774
Undisturbed (in situ) Compression index, $C_c = 0.009$ (LL-10) (from Table 4.1)	0.27	0.27	0.38	0.38
$H \times \log (p_2 / p_1) \div (1 + e_0)$	24 × log (2197 / 768) ÷ (1 + 0.64) = 6.680	36 × log (2034 / 1088) ÷ (1 + 0.64) = 5.965	48 × log (1732 / 1375) ÷ (1 + 1.00) = 2.406	48 × log (1774 / 1576) ÷ (1 + 0.90) = 1.298
Example Problem 4.4, for normally consolidated soils, using only compression index, C_c, because increased pressure is above the existing overburden pressure.				
Settlement increment due to layer	1.80	1.61	0.91	0.49
Example Problem 4.5, for preconsolidated soils, using only swelling index, C_s, because total increased pressure, p_2, is below preconsolidation pressure.				
Swelling index, C_s (est. 0.15 C_c)	0.04	0.04	0.06	0.06
Unconfined compressive strength, psf	1930	2050	2635	2720
Preconsolidation pressure, p_c, estimated from Eq. 4.2	5350	5670	5960	6160
Settlement increment due to layer, inches	0.27	0.24	0.14	0.08

Answer:
Because the total footing pressure p_2 is less than the estimated preconsolidation pressure for each sub-layer, all soils will be loaded only in the recompression zone of the e-log p curve. If the footing pressure is above the preconsolidation pressure for any one of the sublayers, then follow the procedure illustrated in Problem 4.1. Summarize all values in Table 4.3.

The swelling index C_s is estimated at 0.15 C_c. The settlement of each sublayer is calculated using Equation 4.4 and the total is summed using Equation 4.5. In this example, S = 0.27 + 0.24 + 0.14 + 0.08 = 0.73 in.

4.30 CHAPTER 4

This example problem illustrates why, as discussed in Chapter 8, the governing factor for shallow footings in clays is the compressive strength of the clay. When the allowable foundation pressure is determined by the unconfined compressive strength, the preconsolidation pressure for the clay soils is generally above the increased pressure due to the foundation, and the resulting settlement is typically on the order of one inch or less.

Time Rate of Consolidation

4.6 *Question:*
Plot the data for the load increment of Table 4.4 on the deformation versus log time graph of Figure 4.18.

The loading increment was from 2.0 Tsf to 4.0 Tsf. Calculate the 50 percent time (t_{50}) and determine the coefficient of consolidation c_v and the coefficient of permeability k.

Answer:

a. Using the graphical parabolic procedure, the *zero* dial reading was found to be 159.

b. From the intersection of the primary consolidation curve and the secondary compression curve, the dial reading at 100 percent primary consolidation is 536.

c. The dial reading at 50 percent primary consolidation is at 348 corresponding to t_{50} = 4.2 min.

d. For this sample, $2H = 0.75$ inches. At t_{50}, the height $2H = 0.68$ in. (measured from the initial height, allowing for the incremental ΔRdg.). Then, $H = 0.34$ in.

TABLE 4.4 Data to accompany Example Problem 4.6.

Time, min.	Dial Reading, 0.0001 in.
0.1	188
0.25	202
0.50	219
1.0	245
2.0	281
4.0	337
8.0	406
15.0	463
30	510
60	548
120	570
240	583
480	597
1440	627

e. The coefficient of consolidation c_v is calculated from Equation 4.11:

$$c_v = \frac{0.197\,(0.34)^2}{4.2} = 0.0054 \text{ in}^2/\text{min}$$

4.7 Question:

Assume the deformation versus log time curve of Figure 4.18 represents the clay layer of Figure 4.1; that is, a layer 10 ft thick. Calculate the time to reach (a) 50 percent consolidation and (b) 90 percent consolidation.

Answer:

The time to reach a given degree of consolidation is given by Equation 4.11. c_v was determined in Problem 4.6 to be 0.0054 in²/min. Then,

a. Time for 50% consolidation $= \dfrac{H^2}{c_v} T_v = \dfrac{(5 \cdot 12)^2}{0.0054}\, 0.197 = 131{,}333$ min
$\qquad\qquad\qquad\qquad\qquad\qquad\qquad\qquad\qquad\qquad = 91.2$ days

b. Time for 90% consolidation $= 91.2 \cdot \dfrac{0.848}{0.197} = 392$ days

REFERENCES

1. Casagrande, A. 1936. "The determination of the pre-consolidation load and its practical significance," *Proceedings*, Vol. 3, First International Conference on Soil Mechanics. Cambridge, Mass.
2. Schmertmann, J. 1953. "Estimating the true consolidation behavior of clay from laboratory test results," *Proceedings,* Vol. 79, Separate 311. *American Society of Civil Engineers.* New York.
3. Department of the Army. 1990. "Settlement analysis," *Engineer Manual EM 1110-2-194,* U.S. Army Corps of Engineers, Washington, DC, 30 October (also available as Technical and Engineering Design Guides as Adapted from the US Army Corps of Engineers, No. 9, from the American Society of Civil Engineers, New York).
4. Boussinesq, J. 1883. *Application des potentials a l'etude de l'equilibre et du mouvement des solides elastiques.* Paris: Gauthier-Villars.
5. Westergaard, H.M. 1938. "A Problem of Elasticity Suggested by a Problem in Soil Mechanics: Soft Material Reinforced by Numerous Strong Horizontal Sheets." In *Contribution to the Mechanics of Solids,* Stephen Timoshenko 60th Anniversary Volume. New York: Macmillan.
6. Terzaghi, K. 1954. "Anchored Bulkheads," Transactions, American Society of Civil Engineers. 119: 1243.
7. Teng, W. C. (1962). *Foundation design,* Prentice-Hall, Englewood Cliffs, NJ.
8. NAVFAC. DM 7.01. 1982. "Soil Mechanics," Design Manual DM 7.01, Naval Facilities Engineering Command, Department of the Navy, Alexandria, VA, September
9. Terzaghi, K. 1925. *Erdbaumechanik auf Bodenphysikalischer Grundlage*, Vienna, Deuiticke.

CHAPTER 5
STRENGTH OF SOIL AND ROCK

Soil and rock strength is the maximum resistance that can be developed to external and internal stresses tending to cause *failure*, or excessive deformation or rupture. Soils and rocks have a lower tensile strength than compressive strength. In compression, the normal mode of failure is in shear. The shear strength of soil and rock is developed from a combination of sliding friction between particles, cohesion and adhesion between soil particles, interlocking of solid particles, and grain crushing. Shear strength is reduced by any pore-water pressure developed during particle movement.

The shear strength of a soil or rock is the governing factor in several types of geotechnical engineering problems such as:

1. Slope stability, including natural hillsides, cuts, embankments, and earth dams
2. Ultimate bearing capacity of soils
3. Lateral pressure against retaining walls, sheeting, or bracing
4. Skin friction developed by piles.

COMBINED STRESSES

Although the settlement of soils can be described by simple cases of one-dimensional stress and compression, many other problems of soil strength and deformation, such as those given above, involve three-dimensional stresses. A *stress* is defined as a force per unit of area. A stress applied at some angle to a plane surface can be resolved into two components: a stress perpendicular to the surface, known as the *normal* stress, and a stress acting in the plane of (parallel to) the surface, known as the *shear* stress.

Consider the stresses acting on the elemental cube shown in Figure 5.1(a). The principles of solid mechanics demonstrate that there are three independent, perpendicular, principal stresses acting on the three perpendicular, principal planes. The largest of the three principal stresses is the *major principal* stress, σ_1. The *intermediate principal* stress is σ_2 and the smallest is the *minor principal* stress, σ_3. In soil mechanics, it is common to work with a two-dimensional system consisting of the major and minor principal stresses, σ_1 and σ_3.

As shown in Figure 5.1(b), the shear and normal force on a plane making an angle θ can be determined by the laws of statics. Thus, the normal stress on the plane is

$$\sigma_\theta = \frac{\sigma_1 + \sigma_3}{2} + \frac{\sigma_1 - \sigma_3}{2} \cos 2\theta \tag{5.1}$$

and the shear stress parallel to the plane is

$$\tau_\theta = \frac{\sigma_1 - \sigma_3}{2} \sin 2\theta \tag{5.2}$$

MOHR'S CIRCLE OF STRESS

A graphical procedure for solving Equations 5.1 and 5.2 was developed by Mohr [1]. The procedure solves for the normal and shear stresses on the plane shown in Figure 5.1 for all values of angle θ. Combining Equations 5.1. and 5.2 appropriately results in an equation of a circle of the form

$$\left(\frac{\sigma_1 - \sigma_3}{2}\right)^2$$

The center of the circle always lies at a point where $\tau = 0$ and $\sigma = [(\sigma_1 + \sigma_3)/2]$ on the σ_θ-axis and whose radius is $[(\sigma_1 - \sigma_3)/2]$.

The generic Mohr's circle in Figure 5.2 shows the normally used sign convention. Principal stresses are plotted on the x-axis, with compressive (positive) stresses to the right and tensile (negative) stresses to the left of the origin. Positive shear stresses are plotted upward if they produce a counterclockwise couple, and are plotted downward if the couple is clockwise. Therefore, the coordinates of a point (σ,τ) represent the normal and shear

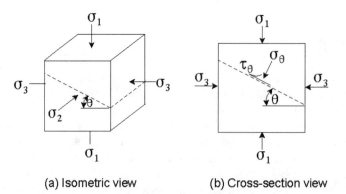

(a) Isometric view   (b) Cross-section view

FIGURE 5.1 Stresses acting on a cube cut by a plane perpendicular to plane of σ_2 and making an angle of θ with the plane of σ_1.

forces on a plane at an angle θ measured counterclockwise positive as shown. In a three-dimensional system, the x-axis position of the intermediate principal stress would be intermediate between σ_1 and σ_3.

MOHR-COULOMB FAILURE CRITERIA

Coulomb [2] empirically showed that the shear strength of soil could be expressed as

$$s = c + \sigma \tan \phi \tag{5.3}$$

where s = Shearing strength
 c = Cohesion (y-intercept)
 σ = Normal force on the critical failure plane
 ϕ = Angle of internal friction (slope indicating rate of change of shearing resistance with change in normal force).

Mohr [1] presented a theory for failure that applies well to soils, rock, concrete, and similar materials. According to this theory, a material fails along the plane and at the time at which the angle between the resultant of the normal and shearing stresses and the normal stress is at a maximum that is, when the combination of normal and shearing stresses produces the maximum obliquity angle α_m, as shown in Figure 5.2.

The combination of the Mohr circle and Mohr failure theory with the Coulomb shear strength equation leads to the Mohr-Coulomb failure theory, shown in Figure 5.3. Only the upper half of the circle is normally shown because of the symmetry of the Mohr circle and because, in general, only compressive stresses are involved. Failure occurs on a plane

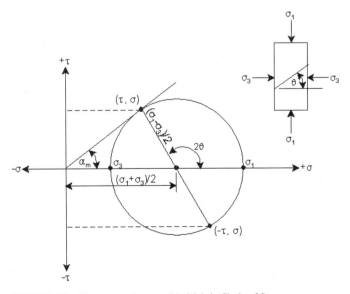

FIGURE 5.2 Sign conventions used in Mohr's Circle of Stress.

making an angle θ with the horizontal when the shearing stress on that plane reaches the Coulomb shearing strength criterion at point A. The Mohr obliquity angle α_m shown in Figure 5.2 is then equal to the Coulomb friction angle ϕ and

$$\theta = 45° + \frac{\phi}{2} \tag{5.4}$$

From the geometry of Figure 5.3 it can be seen that

$$\sin \phi = \frac{\frac{(\sigma_1 - \sigma_3)}{2}}{\frac{(\sigma_1 + \sigma_3)}{2} + \frac{c}{\tan \phi}} \tag{5.5}$$

Combining Equation 5.4 with Equations 5.1, 5.2, and 5.3 it can be shown that

$$\sigma_1 = \sigma_3 \cdot \tan^2\left(45 + \frac{\phi}{2}\right) + 2c \cdot \tan\left(45 + \frac{\phi}{2}\right) \tag{5.6}$$

In a saturated soil, the total normal stress at any point is the sum of the effective intergranular stress and the pore-water pressure, that is, $\sigma = \sigma' + u$. The effective stress σ' is carried by the soil solids and u is carried by the water. Recognizing this, Terzaghi [3] rewrote Equation 5.3 as

$$s = c + (\sigma - u) \tan \phi = c + \sigma' \tan \phi' \tag{5.7}$$

Typically, the value of the cohesion c is zero for cohesionless soils such as clean, free-flowing sand and inorganic silt. For normally consolidated cohesive soils, including all clayey soils, the cohesion is approximately zero and for overconsolidated clayey soils the cohesion is greater than zero. As will be discussed below, partial saturation creates an *apparent* cohesion.

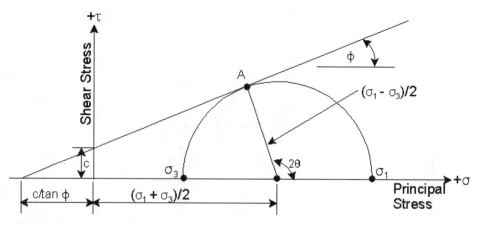

FIGURE 5.3 Mohr-Coulomb failure criterion.

TESTS FOR SHEAR STRENGTH OF SOILS

The shear strength parameters of a soil can be determined in the laboratory. Of the many types of tests that have been tried with varying success, two principal test methods are in common use—the direct shear test and the triaxial compression test.

Direct Shear Test

A thin sample of soil is placed in a container that holds a metal box with a moveable upper and a fixed lower half, as shown in Figure 5.4. Samples are typically 0.5 in. thick by 2.5 in. in diameter (1.27 cm by 6.35 cm), but they are sometimes larger. Porous stones may be placed under and on top of the sample unless no drainage of pore water is to be allowed during the test, in which case they are replaced with solid disks. The initial void ratio of the sample is recorded. A dial indicator or other motion-measuring device is placed on the upper plate to measure vertical motion of the sample during shear, and another is placed against the upper half of the box to measure the horizontal shear displacement.

In the test, the sample is saturated by filling the container surrounding the shear box with water. A normal force is applied to the sample and the sample is allowed to consolidate to at least 100 percent primary (except in the unconsolidated-undrained test). Then a horizontal shearing force is applied to the upper box and increased at a constant rate of deformation until shear failure occurs. A graph may be made showing two relationships: shear force versus horizontal deformation and vertical deformation versus horizontal deformation.

The maximum shearing resistance is plotted as a point on a graph of shear strength versus normal force. Additional samples at the same initial void ratio are tested at different normal force magnitudes. At least three samples are usually tested to develop a straight line to fit the Coulomb shear strength relationship, Equation 5.3.

Triaxial Shear Test

A cylindrical sample of soil is encased in an impervious flexible (usually rubber) membrane and then placed in a pressure chamber, as shown in Figure 5.5. Porous stones are placed under and on top of the sample to facilitate the drainage of pore water or the measurement of pore-water pressure during the test. Drainage of pore water into and out of the

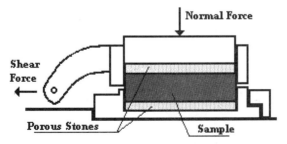

FIGURE 5.4 Schematic of direct shear test device

sample is controlled by drainage valves. The initial void ratio of the sample is recorded. The chamber is filled with a fluid (usually water) and the upper piston placed in contact with the sample. The chamber fluid is pressurized to provide an equal all-around minor principal stress σ_3 on the horizontal and vertical axes of the sample. The piston is used to apply a compression force to the vertical axis of the sample. A dial indicator or other motion-measuring device is attached to the piston to measure vertical deformation. The piston, or deviator ($\sigma_d = \sigma_1 - \sigma_3$), stress is increased gradually and the corresponding deformation is recorded. The ultimate vertical deviator stress $\sigma_{d(max)}$, is added to the constant σ_3 to calculate σ_{1max}.

A Mohr (half-) Circle is then plotted on a Mohr-Coulomb space using the derived σ_{1max} and the applied constant σ_3 as shown in Figure 5.3. Additional samples at the same initial void ratio are tested at different minor principal stress σ_3 magnitudes. At least three samples are usually tested to develop a straight line tangent to all circles to fit the Coulomb shear strength, Equation 5.3.

Types of Laboratory Shear Tests

The major control factors that determine the shearing resistance are the amount of drainage of the pore water permitted during the consolidation phase and during the shearing phase and the rate at which the shearing deformation is applied. These factors are used to simulate pore-water pressure and loading conditions expected to be encountered in the field.

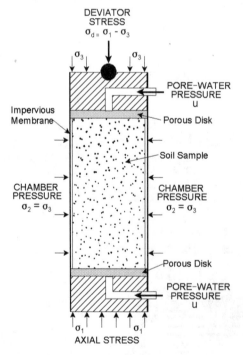

FIGURE 5.5 Schematic arrangement of triaxial shear test. The developed load on the piston is the total axial stress σ_1 which is made up of the deviator stress σ_d, combined with the chamber pressure σ_3

The three major test types are described in Table 5.1.

Consolidated-drained (CD or S) Test. Full drainage is permitted during the *entire* test. The sample is first fully consolidated and then sheared very slowly so that no pore-water pressure is induced. All stresses are effective stresses.

Consolidated-undrained (CU or R) Test. Drainage is permitted only during the consolidation phase. The sample is first fully consolidated and then sheared at such a rate that no further drainage occurs, causing full pore-water pressure to develop due to shear.

Unconsolidated-undrained (UU or Q) Test. No drainage is permitted during the entire test. The sample is not allowed to consolidate. With no drainage permitted, full pore-water pressure is developed due to shear.

Unconfined Compression Test. A cylinder of cohesive soil (one with sufficient clay content and sufficient shear strength to retain its shape) is tested in compression without pre-consolidation and without confining pressure. This is the same as the unconsolidated-undrained triaxial shear test made with an all-around pressure σ_3 equal to zero (atmospheric only).

TABLE 5.1 Direct Shear and Triaxial Shear Test Arrangements

Pressure	CD or S Test	CU or R Test	UU or Q Test
CONSOLIDATION PHASE			
σ_3	Held constant	Held constant	Held constant
σ_1	Equal to σ_3 *	Equal to σ_3 *	Equal to σ_3 *
Drainage	No drainage permitted.	Full drainage permitted.	Full drainage permitted.
	No volume change occurs. Pore-water pressure develops equal to confining pressure.	Pressure maintained until volume change at 100% primary consolidation is reached.	Pressure maintained until volume change at 100% primary consolidation is reached.
SHEARING PHASE			
σ_3	Held constant	Held constant	Held constant
σ_1	Very gradually increased from σ_3 until shear failure occurs.†	Gradually increased from σ_3 until shear failure occurs.†	Gradually increased from σ_3 until shear failure occurs †
Drainage	Full drainage permitted.	No drainage allowed ‡	No drainage allowed ‡
Shearing Rate	Sufficiently slow that. no pore-water pressure will develop. All stresses are effective stresses.	Relatively rapid to very slow. Pore water pressure develops due to shear.	Relatively rapid to very slow. Pore water pressure develops due to shear.

* Consolidating pressure in direct shear is due only to the normal force acting vertically on the shear plane.
† In the direct shear test, force is applied horizontally to the upper block to cause shear on the plane between the upper and lower blocks.
‡ In triaxial shear, pore-water pressure may be measured at the porous disks.

Pore-Water Pressure Resulting from Stress

Pore-water pressure u in a soil element may occur under *undrained* conditions in which water flows into or out of the element. Two undrained conditions that are of interest mainly because of their fundamental significance are pore-water pressure associated with a change in the equal all-around total stress on the element and pore-water pressure resulting from shear deformations. The triaxial shear test is the most widely used means for investigating undrained behavior.

Pore-water Pressure Due to Change in All-around Stress. Skempton [4] defined the change in pore-water pressure Δu resulting from undrained changes in the equal all-around stress $\Delta \sigma$ in terms of a pore-pressure coefficient

$$B = \Delta u / \Delta \sigma_3. \tag{5.8}$$

B equals one for saturated soil and rock materials for which the compressibility of the soil or rock skeleton C is much greater than the compressibility of the solids forming the skeleton C_s. This is the case for all soils as well as for fissured or jointed rock masses. For intact rocks, where C may approach C_s, B is less than one. Such rocks often contain microfissures.

Values of B on the order of $0.5 \pm$ are found in partially saturated soils. In fact, this measurement is used in triaxial testing where back pressure (increased internal water pressure through the drainage system) is used to increase the pore-water pressure in order to cause the air bubbles in the voids to dissolve into the water, thereby effectively causing full saturation. Back pressure is increased, with a corresponding increase in the external confining pressure, until slight changes in σ_3 produce an equal change in u, indicating that B equals one and the sample is fully saturated.

Pore-water Pressure Due to Application of Shear Stress. When the added principal stresses are caused to be unequal, that is, $\Delta \sigma_1 \neq \Delta \sigma_2 \neq \Delta \sigma_3$, the soil is subjected to shear stresses. All soils and rocks tend to experience volume change when subjected to a change in shear stress. Shear-induced pore-water pressures occur under undrained conditions. For saturated soils in a triaxial shear test, where $B = 1$ and $\Delta \sigma_2 = \Delta \sigma_3$, the relationship for the A-coefficient proposed by Skempton [4] is

$$A = \Delta u_d / \Delta \sigma_d \tag{5.9}$$

in which Δu_d and $\Delta \sigma_d$ are the change in pore-water pressure associated with a change in the deviator stress, $(\Delta \sigma_1 - \Delta \sigma_3)$, during shear.

This expression is not used much in practice; however, the general behavior of shear-induced pore-water pressure in saturated soils and the associated A-coefficient are well established. Soils and rocks that tend to contract on application of shear stresses, such as loose sands and normally consolidated clays, develop positive pore-water pressures that lead to positive values of the A-coefficient. Bishop and Henkel [5] showed that A at failure (A_f) is related to the overconsolidation ratio (OCR) of clays, as shown in Figure 5.6.

In contrast, soils and rocks that tend to dilate (expand) on the application of shear stresses develop high, negative pore-water pressures during shear. Dense sands and highly

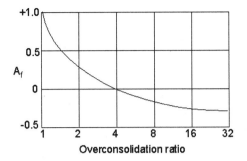

FIGURE 5.6 Change in pore-water pressure during the application of deviator stress. Typical results for normally and over-consolidated clay samples (after Bishop and Henkel [5]).

overconsolidated clays have a negative value of A_f. Values of A are rarely less than about -0.3 because the increased negative pore-water pressure tends to increase the resistance to shear, requiring higher shearing forces.

Relative Density

The shear strength of a clean granular soil is directly related to relative density, and is affected by the angularity of the grains. Relative density is defined as:

$$D_r\% = 100\left(\frac{e - e_{\min}}{e_{\max} - e_{\min}}\right) = 100\left[\frac{\gamma_{d-\max}(\gamma_d - \gamma_{d-\min})}{\gamma_d(\gamma_{d-\max} - \gamma_{d-\min})}\right] \quad (5.10)$$

where D_r = Relative density
  γ_d = Dry unit weight (dry density)
  $\gamma_{d\,\max}$ = Dry unit weight of soil in the densest state that can be attained in a particular laboratory test called *maximum density*
  $\gamma_{d\,\min}$ = Dry unit weight of soil in the loosest state that can be attained in a particular laboratory test called *minimum density*

Definitions

By general agreement among practitioners, the terminology for relative density is
  Loose sand $0 < D_r < 33\%$
  Medium compact sand $33\% < D_r < 67\%$
  Dense sand $67\% < D_r < 100\%$

Laboratory Relative Density Tests

Formal ASTM Standard [6] laboratory tests have been devised for determination of the maximum and minimum densities:

Minimum Density. In accordance with ASTM Standard D4253, the minimum density of a sand is determined by pouring the sand into a container of known volume. The pouring is done through a funnel with a constant, very small drop distance.

Maximum Density. In accordance with ASTM Standard D4254, the maximum density is determined by placing the sand in a container of known volume. The container then is covered with a weight to prevent rebounding, attached to a vibrating table, and then vibrated until no further decrease in volume occurs.

Expedient Field Test. The shaking table for the maximum density test is not always readily available, especially in field laboratories. To overcome this, the Providence District Corps of Engineers, devised an expedient test for determining maximum density. A steel mold similar to that used for compaction testing (Chapter 12) containing the weight-covered sand is tapped soundly on the side with a ball peen hammer until no further decrease in volume occurs.

Effect of Initial Relative Density on Shear Strength

The relationship between the initial relative density of cohesionless materials with the OCR of cohesive soils, and the stress-strain behavior of sands in consolidated-drained tests is shown in Figure 5.7(a). The corresponding volume changes in fully-drained shear are shown in Figure 5.7(b). The pore-water pressures for samples tested in undrained shear with pore-water pressure measurements are shown in Figure 5.7(c)

When a loose sand or normally consolidated clay (NC) is sheared, the shearing resistance increases steadily to a fairly high amount of deformation. Because there is no well-defined peak strength, the shearing resistance at 20 percent strain ($\Delta L/L_o$) is usually used as the ultimate strength. The individual grains or grain pods are initially in an unstable state; shearing deformation causes the grains to move into a denser state comparatively easily because particle interference is not a major factor. This causes the sand or clay to decrease in volume. If the volume change cannot occur due to lack of drainage, then a positive pore-water pressure is developed and the A-coefficient is positive.

For dense sands and high OCR clays, particle interference forms a major part of the shearing resistance. It is necessary in sands, for example, for a grain in a stable state undergoing a shearing deformation to require the sliding-strain energy to climb up and over its neighbor, in addition to the energy to overcome frictional resistance. A similar effect occurs in overconsolidated clays. This causes the sand or clay with full drainage during shear to increase in volume as shown in Figure 5.7(b). The shearing resistance increases with deformation until a peak is reached, after which, in a strain-controlled test, the resistance drops and reaches a fairly constant lower or residual value, nearly equal to that of an initially loose sand or soft clay.

If undrained shear (Q or R) tests are made with pore-water pressure measurement the general trend of the stress-strain curves are the same as shown in Figure 5.7(a)—the higher the sand relative density or the higher the clay OCR, the higher the total stress shearing resistance. The resulting pore-water pressures are shown in Figure 5.7(c) and the *A*-coefficient is negative.

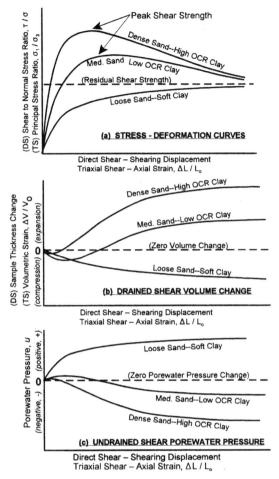

FIGURE 5.7 Shear stress-deformation plots during the shearing phase of typical direct shear (DS) and triaxial shear (TS) tests of dense, medium, and loose sands and of high and medium OCR and normally consolidated clays.

Negative Pore-Water Pressure Due to Partial Saturation

The phenomenon of capillarity can be used to explain the shear-strength behavior of partially saturated soils and of soils that dilate during shearing. Wherever there is an air-water interface, the molecules at the water surface develop a surface energy film known as *surface tension*. If a very small-diameter glass tube is inserted into a container of water, as shown in Figure 5.8(a), the attraction between the glass and the water molecules combined with the surface tension of the water, pull the water up into the tube to a height h above the water level. The upper surface of the water, the *meniscus*, joins the walls of the tube at an angle known as the *contact angle*. The pressure in the water in the tube above the surface in the container is negative and is being held in place by the surface tension of the water. The height of rise h is a function of the type of material and the diameter of the tube.

Because the radius of the tube is in the denominator of the equation shown in Figure 5.8(a), the larger the diameter of the tube, the lower the surface tension and, therefore, the lower the capillary rise.

The voids in a soil have very small, variable diameters. Wherever a water-air combination exists in a partially saturated soil's void space, surface tension will act on the soil particles, pulling them together with a force P known as the *contact pressure*, as shown in Figure 5.8(b). The contact pressure acts as an additional normal force on the shear plane. The effect is as if the soil has a certain amount of cohesion, shown by the intercept on the Mohr-Coulomb plot shown in Figure 5.3. This strength is known as *apparent cohesion* and its value obviously varies with the degree of saturation, changing the surface tension and disappearing entirely when the soil becomes saturated.

SHEAR STRENGTH RELATIONSHIPS

The behavior of soils and rocks undergoing stress in a three-dimensional stress field in actual field situations can be demonstrated by laboratory direct shear and triaxial shear tests.

Consolidated-Drained (S-Test) Shear of Saturated Soils

The coefficient of permeability of clean sand is relatively high. Therefore, at a normal loading rate during a fully drained test, the excess pore-water pressure developed due to consolidation and the shearing stress is quickly dissipated, so that essentially full drainage conditions exist throughout and all stresses are effective stresses. The friction angle developed by plotting the peak shear strength (Figure 5.7) versus normal load is shown in Figure 5.9 for a dense sand and a loose sand. The effective stress friction angle for a granular material is directly related to its relative density and grain size distribution, as shown in Figure 5.10, which may be used for a preliminary approximation of the friction angle but should not be used without test verification.

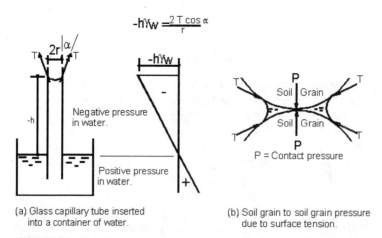

(a) Glass capillary tube inserted into a container of water.

(b) Soil grain to soil grain pressure due to surface tension.

FIGURE 5.8 Negative pore-water pressure from capillarity caused by partial saturation.

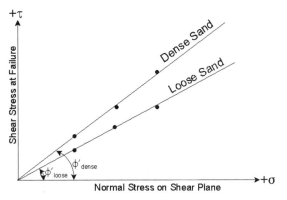

FIGURE 5.9 Direct shear test results for a dry or a saturated clean sand at two relative densities.

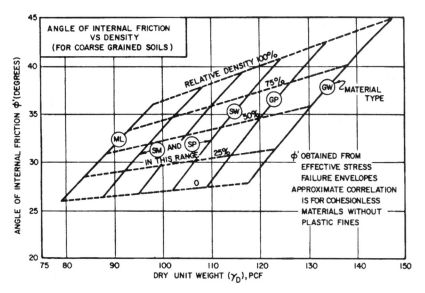

FIGURE 5.10 Approximate correlation of drained friction angle of coarse-grained materials with material type and relative density [7]).

The coefficient of permeability of clayey soils is relatively low. Therefore, for a fully drained test, excess pore-water pressure developed during the consolidation phase must first be fully dissipated by consolidating to 100 percent of primary. Pore-water pressures due to shearing stresses must be continually dissipated so that essentially full drainage conditions exist and all stresses are effective stresses. This is accomplished by applying the shearing deformation very slowly, at a rate of deformation in a strain-controlled test determined by the permeability of the sample.

The result of a series of very slow, fully drained, direct shear tests is shown in Figure 5.11(a) and of triaxial shear tests in Figure 12(a). For a normally consolidated clay, the void ratio changes with consolidating pressure in accordance with Equation 4.1 and as shown in Figure 4.3. The maximum shearing resistance, τ_{max}, increases directly with the void ratio change, forming a straight line relationship that satisfies the Coulomb equation, Equation 5.3, but with zero cohesion. This is consistent with the development of shearing strength of a clayey soil sedimented from a slurry and consolidated under its own weight.

For an over-consolidated clay, if the normal pressure on the shear plane due to the vertical force in direct shear or the equal all-around pressure σ_3 in triaxial shear is lower than the pre-consolidation pressure, there will be only a very small volume change during the consolidation phase. This volume change is in the recompression zone of Figure 4.3 and is smaller than one that would have been achieved if the sample had been normally consolidated at the same all-around pressure. Therefore, the shear strength will be higher than that for a corresponding normally consolidated sample at the same normal pressure. This is shown in Figure 5.12(a) as a circle with principal stresses marked with the subscript o.

The slope of the curve in this region will exhibit a *cohesion* intercept and a flatter *friction angle* than that in the normally consolidated region at a higher normal force or higher all-around pressure. The $\tau - \sigma$ curve represents the rate at which the maximum shearing resistance varies with normal force on the shear plane.

When the normal pressure on the shear plane is greater than the preconsolidation pressure, the sample is in the virgin compression zone of the pressure-void ratio curve of Figure 4.3 and the sample behaves as a normally consolidated clay which, in fact, it has become under the higher consolidating pressure.

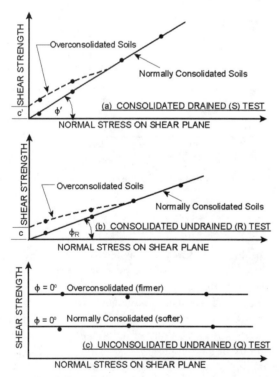

FIGURE 5.11 Direct shear test results for saturated clayey soils in the S-, R-, and Q-test modes.

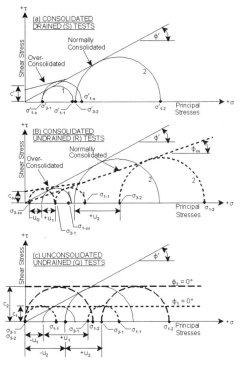

FIGURE 5.12 Triaxial shear test results for saturated clayey soils in the S-, R-, and Q-test modes.

Consolidated-Undrained (R-Test) Shear of Saturated Soils

The result of a series of consolidated-undrained direct shear tests is shown in Figure 5.11(b) and of triaxial shear tests in Figure 5-12(b). In Figure 5.12, effective stress Mohr circles are shown as solid and total stress circles are shown as dashed lines. It can be shown that, although the major and minor principal stresses are not the same as in the consolidated-drained (S) test, the principal stresses are related as

$$\sigma_1 - \sigma_3 = \sigma_1' - \sigma_3' \qquad (5.8)$$

This indicates that although the radius of the Mohr circle, $\sigma_1 - \sigma_3/2$, remains the same for both total stress and effective stress circles, the effective stress circle is displaced laterally by an amount equal to the pore-water pressure.

For loose sands and normally consolidated clays, because no volume change is permitted during shear, a *positive* pore-water pressure, $+u$, is developed during shear as shown in Figure 5.7(c). At a constant total stress-confining pressure (the normal force in the direct shear or the all around-pressure, σ_3 in the triaxial test) the effective stress on the shear plane is developed according to Equation 5.7. Failure occurs when the effective shear stress on the critical shear plane becomes tangent to the Coulomb strength envelope. This is shown in Figure 5.12(b), in which the two total stress circles, numbered 1 and 2, for normally consolidated clays are of the same size as the corresponding effective stress circles, but are displaced by positive pore-water pressure. The slope of the Coulomb shear-strength line in total stress terms, ϕ_{cu}, is smaller than the effective stress shear strength angle, ϕ'.

If a dense sand or an over-consolidated clay sample with a normal pressure on the critical shear plane lower than the preconsolidation pressure is tested, then the discussion above for the S-test applies equally to the R-test. During the shearing phase, a dense sand or an over-consolidated clay will try to dilate as shown in Figure 5.7(c). Since no drainage is permitted, this creates a *negative* pore-water pressure, $-u$, and the Mohr circle is displaced to the right. This is shown in Figure 5.12 (b) as a circle with principal stresses marked with the subscript o. As with the S-test, once the magnitude of the confining pressure becomes greater than the pre-consolidation pressure of the sample, the material behaves as a normally consolidated material, which in fact it is.

Unconsolidated-Undrained (Q-Test) Shear of Saturated Soils

When a series of saturated samples (with B-coefficient = 1) that have the same initial void ratio are tested at different confining pressures, each will develop an initial pore-water pressure equal to the confining pressure if no drainage is permitted during the application of the consolidating pressure. With no change in volume, all samples will have the same effective stress-shearing resistance and all Mohr circles, according to Equation 5.8, will have the same diameter, as shown for samples 1 and 2 in Figure 12(c). The resulting total stress envelope for all samples of the same initial void ratio will be a horizontal line, as seen in Figures 11(c) and 12(c), indicating a total stress friction angle equal to zero and a total stress intercept that is interpreted as *cohesion*.

During the early days of soil testing, before the effective stress concept of Equation 5.7 became universally known and understood, clayey soils exhibiting this behavior were considered to have $\phi = 0\%$ and cohesion and therefore were called cohesive soils. And because clean sands and gravels generally were tested in the fully drained condition and did not exhibit a cohesion intercept, they were invariably referred to as cohesionless soils.

Figures 11(c) and 12(c) both show two levels of cohesion that are due to differing levels of initial void ratio. The denser the soil, whether sand or clay, the greater the Q-test, or undrained strength. The Q-test, however, is rarely run because it does not directly account for differences in the initial void ratio.

Unconfined Compression Test of Saturated Clays

The unconfined compression test is a special case of the unconsolidated-undrained triaxial shear test. This is one of the most widely used tests for measurement of the compressive strength of clayey soils, mainly because both the testing equipment and the test procedure are simple. It provides practically the same results as a more complicated Q-triaxial test. Its primary applicability will be discussed below.

At failure, the minor principal stress, σ_3, is zero. Therefore, in total stresses, the shear strength at failure is $\sigma_1/2$, which is equal to the cohesion, c_u. The act of sampling relieves the *in situ* lateral pressure on the sample. Since there is no confining pressure to replace the *in situ* stress, the sample attempts to expand, creating a negative pore-water pressure, $-u$. Thus, as shown in Figure 5.12(c), the sum of the positive axial stress, σ_1 with negative pore-water pressure, $-u$, leads to a critical effective stress at failure.

Shear Strength of Partially Saturated Soils

The tension (negative pore-water pressure) in the water in a partially saturated soil causes an internal effective pressure on the grain-to-grain contact that acts in the same manner as an external force. This concept is particularly suited to understanding the shear strength of compacted soils and soil samples taken from above the water table, all of which are invariably partially saturated.

As shown in Figure 5.8, the surface tension due to partial saturation increases the apparent strength of the soil by increasing intergranular pressure. The strength will vary as the degree of saturation varies. The preconsolidation pressure, Figure 4.3, is also increased along with strength and will vary as the degree of saturation varies. The permeability of a soil varies with the degree of saturation because the air bubbles in the voids impede the flow of water through the soil.

The Terzaghi-Coulomb shear-strength equation, Equation 5.7, indicates that partial saturation of granular soil, resulting in a negative pore-water pressure, causes an increase in the intergranular pressure which creates a higher shearing resistance for the same normal pressure on the shear plane. This is illustrated in Figure 5.13, which shows that a partially saturated sandy soil has an apparent cohesion. This explains why it is possible to build sand castles with very little or no normal pressure on the grains and why the castles collapse when they become saturated with water.

The clay particles in a cohesive soil tend to form flocs and behave somewhat like discrete particles that react to partial saturation in about the same manner as granular soils. Fredlund [8] revised the Terzaghi-Coulomb equation to account for the pressure in the void air and in the pore-water:

$$s = c + (\sigma_n - u_a) \tan \phi + (u_a - u_w) \tan \phi_b \tag{5.9}$$

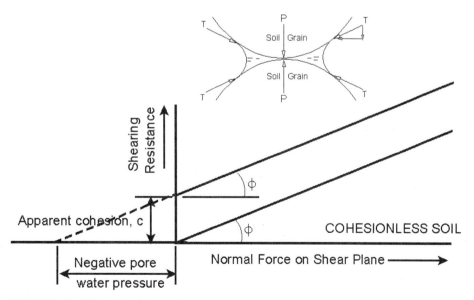

FIGURE 5.13 Effect of partial saturation on the shear strength of a cohesionless, sandy soil.

where σ_n = Normal stress on shear plane
  u_a = Pore-air pressure
  u_w = Pore-water pressure
  ϕ = Angle indicating the rate of increase in shear strength with respect to net normal stress, $(\sigma_n - u_a)$
  ϕ_b = Angle indicating the rate of increase in shear strength due to a change in the matric suction, $(u_a - u_w)$.

Because the permeability of clay is very low, there is no drainage during an unconsolidated-undrained (Q) test and the shear strength is constant ($\phi = 0$), for any given initial void ratio, with respect to external total stress loading. However, the part of the shear strength due to partial saturation increases with a decrease in the degree of saturation (increase in matric suction), as shown in Figure 5.14. Therefore, the shear strength of the partially saturated cohesive soil varies as the degree of saturation.

Sensitivity of Clays to Disturbance

For clayey soils deposited in many natural environments, it is possible to completely remold a cohesive soil sample, destroying its natural structure, and then re-form it to its original volume, shape, and water content. An unconfined compression test can then be used to determine its remolded strength. *Sensitivity* is the ratio of the unconfined compressive strength of an undisturbed cohesive soil to the unconfined compressive strength of the same soil sample whose structure has been destroyed by thorough remolding, or

$$S_t = \frac{\text{Undisturbed } q_u}{\text{Remolded } q_u} \qquad (5.11)$$

A comparison of typical stress-strain curves from strain-controlled laboratory unconfined compression tests is shown in Figure 5.15. When there is no well-defined peak strength, the ultimate strength normally is taken as the shearing resistance at 20 percent strain (some organizations use 15 percent). At a high value of strain, the shearing itself causes considerable remolding and reorientation of the structure of the clay, causing the shearing resistance to decrease and approach the shearing resistance of an initially remolded sample.

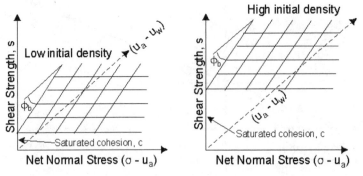

FIGURE 5.14 Three-dimensional representation of the undrained shear strength of a partially saturated clayey soil, ilustrating the presence of an apparent cohesion due to surface tension [8].

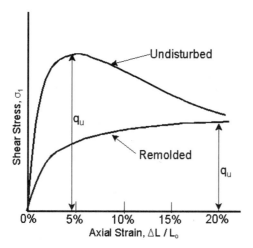

FIGURE 5.15 Unconfined compression stress-strain curves for undisturbed and remolded samples of clayey soil.

The loss of strength on remolding, for most clays, is primarily due to the reorientation of the randomly oriented, or flocculated, clay platelet structure to a parallel, or dispersed, orientation. The flocculated structure is mostly associated with deposition of the clay in a salt-water environment in which salt ions affect the edge-to-face attraction of the platelets.

Sensitivity indicates the strength available at a high deformation and at a high amount of strain, as would occur after a landslide has started. Because of differences in original grain structure, flocculated or dispersed, and in mineralogy, different cohesive soils will have different values of sensitivity. For any given saturated clay sample, the shear strength varies directly with density and, therefore, with water content, porosity, and void ratio. The liquidity index varies directly with water content for a saturated, remolded soil, relative to the liquid- and plastic-limits water contents, and is directly related to the shear strength for soils of low sensitivity. The liquidity index serves well as a quality check on undrained shear-strength tests of low sensitivity soils.

The sensitivity ratio, S_t, of most clays ranges from about 1 to more than 6. The sensitivity ratio of highly flocculent marine clays that have been raised tectonically above sea level and had their salt ions leached can reach between 10 and 80. Such clays are referred to as *quick* clays. Deposits of quick clays have been found in northern North America and in Scandinavia.

Partial remolding of clays occurs in the tube sampling of clays from bore holes. This topic will be discussed in a later chapter under the heading of soil sampling. If the thickness of the sampling tube is too great, the effect with be similar to squeezing toothpaste from a tube. That is, the soil will be squeezed into a tube of smaller diameter than the sample because of the thickness of the tube walls. This, obviously, will cause some remolding.

Selection of Shear Strength for Design

It is evident from the above discussion that the shear strength of any soil depends on its grain-size distribution, initial void ratio, and on the pore-water pressure at failure. Excess pore-water pressures may be caused by soils that tend to compress during shear, whereas dilating soils cause negative pore-water pressures. Excess pore-water pressures, whether positive or negative, will tend to dissipate by drainage, which is a function of the permeability of the soil and of the rate at which the shearing stresses are applied.

Sands and Gravels. Clean sands and gravels, with few or no cohesive fines, will have a relatively high permeability. The drainage rate under most loading rates will be so great that excess pore-water pressures will not be able to form and all stresses will be effective stresses. Then, the effective stress parameters can be determined by a consolidated-drained (S) test, either triaxial or direct shear. The major difficulty is the fact (discussed in a later chapter) that *in situ* undisturbed samples of sand or gravel are almost impossible to obtain. It is thus necessary, in the laboratory, to re-form samples in the test device at the *in situ* void ratio.

Soft to Firm Clays. For soft to firm clayey soils, the lowest factor of safety normally occurs at the end of construction. At that time the full total shear force is first applied to the soil mass. With time, the pore-water pressure developed as a result of shearing stresses is dissipated by migration to surrounding soil having a lower pore-water pressure. The soil becomes stronger with time, increasing the factor of safety.

For those cases, such as shallow footings, where the increase in strength with depth as measured in the R-test, is insignificant, the unconfined compression test is satisfactory. This easy-to-run test represents the end-of-construction shear strength, and the expense of a more involved R-test is generally unnecessary.

When there is a significant increase in total stress-shear strength within a depth that will be affected by the load, however, one or more R-tests will indicate the probable increase with overburden. This occurs when analyzing pile skin friction or the stability of a slope.

Stiff Clays. The apparent high shear strength of stiff clays is not always permanent. The negative pore-water pressure developed in a stiff, highly over-consolidated clay during shear explains why slopes cut in such a clay often fail after several years, even though an analysis using total stresses indicates the slope is safe. The negative pore-water pressure developed due to a shear stress caused by removal of lateral soil support to form the slope creates a suction force. After a length of time dependent on the permeability of the clay and access to the water table, the soil suction eventually will cause an increase in the water content of the clay, which leads to a general softening and failure. Therefore, an effective stress analysis with consideration of the potential for strength change in time should be used.

EXAMPLE PROBLEMS

5.1 Question:

A compressive stress of 100 kPa (1.04 tons/ft^2) is applied axially (perpendicularly) to the upper surface of a cube of material whose height is twice its width, as shown in Figure 5.1. The cube is subjected to an equal all-around horizontal pressure ($\sigma_2 = \sigma_3$) of 50 kPa (0.52 tons/ft^2). Find the normal and shear forces on a plane making an angle of 60° with the horizontal. (*Hint*: solve using Equations 5.1 and 5.2 and then solve using Mohr's Circle as shown in Figure 5.2).

Solution:

$\sigma_1 = 100$ kPa and $\sigma_3 = 50$ kPa

Equation 5.1 $\qquad \sigma_\theta = \dfrac{100 + 50}{2} + \dfrac{100 - 50}{2} \cos 2\,(60) = 62.5$ kPa

Equation 5.2 $\qquad \tau = \dfrac{100 - 50}{2} \sin 2\,(60) = 21.7$ kPa

The graphical solution to this problem, using Mohr's Circle, is shown in Figure 5.16.

5.2 Question.

A series of consolidated-drained direct shear tests at the same initial void ratio was made on specimens taken from a sample of saturated, silty clay. The test results are:

Test Number	1	2	3	4
Normal stress on shear plane, σ, kPa	50	100	200	300
Shear stress at failure, τ_f, kPa	52.1	70.6	126.2	187.8

The silty clay is overconsolidated. Plot the test data on a $\tau - \sigma$ graph. Find the shear strength parameters (a) in the over-consolidation and in (b) the normal consolidation ranges, and (c) find the soil's preconsolidation pressure.

Solution:

The shear stresses, τ_f from the individual tests are plotted against normal stresses, σ, in Figure 5.17. The plot indicates that (a) at normal loads below the preconsolidation pressure, $c = 40$ kPa and $\phi = 20°$, and (b) at normal loads above the preconsolidation pressure, $c = 0$ and $\phi = 32°$, and, finally in (c) the pre-consolidation pressure is 120 kPa.

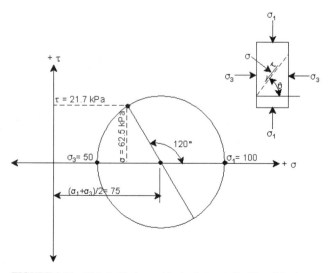

FIGURE 5.16 Mohr's Circle graphical solution to Problem 5.1.

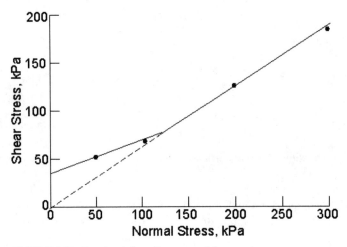

FIGURE 5.17 Results of direct shear tests of dune sand.

5.3 Question
A consolidated-drained (S) test was made on two sections of a tube sample of saturated, normally consolidated clay having the same initial void ratio. The results were as follows:

Sample No.	1	2
σ_3 =	2500 lb/ft²	5000 lb/ft²
$(\Delta \sigma_d)_f$ =	4400 lb/ft²	8800 lb/ft²

Determine (a) the angle of internal friction, ϕ, (b) the angle, θ, that the failure plane makes with the major principal plane, and (c) the normal stress and shear stress on the failure plane.

Solution:
For a normally consolidated soil, the Coulomb failure equation is $\tau_f = \sigma' \tan \phi'$.

The effective major effective stress for sample No. 1 is $\sigma_1 = 2500 + 4400 = 6900$ lb/ft².
The effective major effective stress for sample No. 2 is $\sigma_1 = 5000 + 8800 = 13{,}800$ lb/ft².

(a) The Mohr's circles and failure envelope are shown in Figure 5.18. The friction angle may be determined graphically or by using Equation 5.5:

$$\sin \phi = \frac{\sigma'_1 - \sigma'_3}{\sigma'_1 + \sigma'_3} = \frac{6900 - 2500}{6900 + 2500} = \frac{13800 - 5000}{13800 + 5000} = 0.468$$

$$\phi = 27.9°$$

(b) Using Equation 5.4, solve for θ.

$$\theta = 45 + \frac{\phi}{2} = 45 + \frac{27.9}{2} = 59°$$

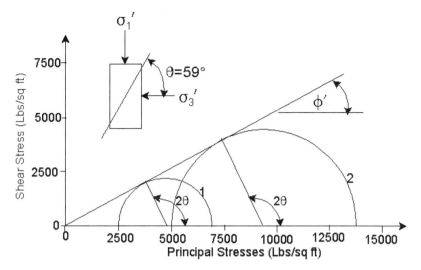

FIGURE 5.18 Effective stress envelope for consolidated-drained tests of a normally consolidated clay.

(c) Using Equations 5.1 and 5.2, solve for τ_f and σ'.

$$\sigma' = \frac{6900 + 2500}{2} + \frac{6900 - 2500}{2} \cos 2(59) = 3667 \text{ psf}$$

$$\tau_f = \frac{\sigma'_1 - \sigma'_3}{2} \sin 2\theta = \frac{6900 - 2500}{2} \sin 2(59) = 1942 \text{ psf}$$

5.4 Question:
A sample of saturated, medium sand is consolidated in the triaxial cell under an equal all-around pressure of 30 lb/in². Then the drainage valves are closed and the cell pressure is increased to 100 lb/in². A pore-water pressure of 70 lb/in² was measured. Maintaining the cell pressure, the axial stress is increased, measuring at failure $(\Delta \sigma_d)_f = 20$ lb/in², with the increase in pore-water pressure measuring at $\Delta u = 19$ lb/in².

A second specimen of the same sand at the same initial void ratio is consolidated under an equal all-around pressure of 10 lbs/in², after which the drainage valves are left open and the axial stress is increased to failure under fully drained conditions.

For the first specimen,
(a) What was the Skempton *B*-parameter during the consolidation phase and what does this indicate?
(b) What was the Skempton *A*-parameter measured during the shearing phase and what does this indicate?
(c) What is the R-test friction angle?
(d) What is the drained friction angle?

For the second specimen,
(e) What is the value of the total axial stress, σ'_1, at failure?
(f) What is the drained shear strength?

Solutions:

(a) Skempton's pore-pressure coefficient $B = \Delta u / \Delta \sigma_3$ (Equation 5.8). For the first specimen, after the drainage valves were closed, the pore-water pressure, Δu, increased to 70 lb/in² for an equal all-around pressure increase, $\Delta \sigma_3$, of 70 lb/in², for $B = 1$. This indicates that the sample was fully saturated.

(b) Skempton's pore-pressure coefficient $A = \Delta u_d / \Delta \sigma_d$ (Equation 5.9). For the first specimen, the shear strength and pore-water pressure at failure give $A = 19/20 = 0.95$. Referring to Figure 5.6, it is seen that the sand is very loose.

(c) Using the total stresses at failure, $\sigma_1 = 120$ psi and $\sigma_3 = 100$ psi, and using Equation 5.5,

$$\sin \phi = \frac{120 - 100}{120 + 100} = 0.091$$

$$\phi_{cu} = 5.2°$$

(d) Using the effective stresses at failure, $u = 89$ psi, $\sigma_1 = 31$ psi, and $\sigma_3 = 11$ psi, and using Equation 5.5,

$$\sin \phi = \frac{31 - 11}{31 + 11} = 0.476$$

$$\phi' = 28.4°$$

(e) For the second specimen, tested under fully undrained conditions and having the same drained friction angle as the first specimen, the lateral effective stress, $\sigma_3 = 10$ psi. Then, using Equation 5.6,

$$\sigma_1 = 10 \tan^2 (45 + 28.4 / 2) = 28.1 \text{ psi}$$

(f) As shown by Equation 5.8, both the drained and the undrained shear strength are the difference between the major and minor principal stress at failure. Thus

$$(\sigma'_1 - \sigma'_3)_f = 28.1 - 10 = 18.1 \text{ psi}$$

REFERENCES

1. Mohr, O. 1900. "Welche umstände bedingen die elastizitätsgrenze und den bruch eines materiales?," *Zeitschrift des Vereines Deutscher Ingenieure*, vol. 44, pp. 1524-1530, 1572-1577.

2. Coulomb, C. A. 1776. "Essai sur une Application des Règles des Maximis et Minimis à quelques Problèmes de Statique Relatifs à l'Architecture" (An Attempt to apply the rules of maxima and minima to several problems of stability related to architecture), *Mem. Acad. Roy. des Sciences, Paris,* vol. 7, pp. 343-382.

3. Terzaghi, K. 1936. "The shearing resistance of saturated soils and the angles between the planes of shear," *Proceedings, 1st International Conference on Soil Mechanics and Foundation Engineering*, Cambridge, Mass., vol. 1, pp. 54-56.

4. Skempton, A. W. 1954. "The pore-pressure coefficients A and B," *Geotechnique*, vol. 4, pp. 143-147.

5. Bishop, A. W. and Henkel, D. J. 1957. *The measurement of soil properties in the triaxial test*, Edward Arnold, London.

6. ASTM. 1999. "Natural building stones; soil and rock; geotextiles," *1999 Annual Book of ASTM Standards*, vol. 04.08, American Society for Testing and Materials, Philadelphia, Penn.

7. NAVFAC DM 7.01. 1982. *Soil Mechanics, Design Manual 7.01*. U. S. Department of the Navy, Naval Facilities Engineering Command, Alexandria, VA, September.

8. Fredlund, D. G. 1997. "An introduction to unsaturated soil mechanics," *Unsaturated Soil Engineering Practice, Geotechnical Special Publication No. 68*, American Society of Civil Engineers, New York.

CHAPTER 6
ENGINEERING GEOLOGY OF ROCKS AND SOILS

ORIGIN OF ROCKS AND SOILS

The ultimate source of all rock and soil is the molten magma from deep in the earth. The minerals are nearly all forms of silicates with various other metallic elements. The *rock cycle* starts with the hardened magma, *igneous* rock. The rock mass can be changed by a combination of extreme pressure, heat, distortion, and solution to form *metamorphic* rock. Weathering of the surface of the rock, whether igneous or metamorphic, can be followed by the transportation of the weathered particles to a new location and subsequent induration into *sedimentary* rocks.

Igneous Rocks

The two primary forms of igneous structure are *intrusive* or *plutonic* and *extrusive*. Intrusive rocks form as the magma cools slowly below the surface, resulting in coarse-grained *granite* and granite-like masses, rich in quartz. Extrusive rocks form as the magma cools rapidly at or near the surface, resulting in *basalt* and other rocks rich in ferromagnesian minerals. Magma ejected from volcanos forms as lava beds and ash beds.

Sedimentary Rocks

The physical and chemical weathering processes (discussed below) acting on the rock create particles small enough to be transported by water, wind, glacial action, or gravity, where they are sedimented at a new location. Pressure and/or cementation harden the sediments into sandstone, siltstone, conglomerate, shales, or limestone (calcium carbonate) and dolomite (magnesium carbonate).

Metamorphic Rocks

A combination of heat, pressure, solution, and distortion alter the minerals in the original rock into new forms. Granites become *gneiss*. Basalt becomes *schist*. Sandstones and siltstones are altered into *quartzites*. Shales are altered into *slates* or *mica schists*. And limestones and dolomites are altered into *marble*.

SOIL FORMING PROCESSES

Soil has different meanings in different fields. To a *pedologist*, soil is the mineral and organic substance existing at the upper few feet of the earth's surface, which grows and develops plant life, the *A* and *B* horizons shown in Figure 6.1. To a *geologist*, soil is the mineral and organic material in the relatively thin surface zone within which roots occur (the same as the pedologist's definition). The rest of the earth's crust is grouped under the term *rock* irrespective of its hardness or degree of induration. To an *engineer*, soil consists of the sediments and other unconsolidated accumulations of solid particles produced by the mechanical and chemical disintegration of rocks.

Weathering Processes

Weathering produces engineering soil by breaking the native rock into discrete particles. Weathering occurs due to mechanical (physical) and chemical-biological processes. The surface of the soil layer is weathered at an accelerated rate because of plant and animal activity, forming pedologic soil that will support plant life. The upper few feet of a typical temperate zone soil profile are shown in Figure 6.1.

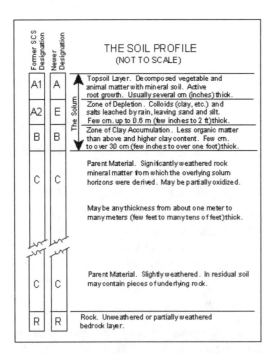

FIGURE 6.1 Upper few feet of a typical temperate zone soil profile showing plant growth zone, engineering soil, and rock.

Mechanical Weathering (Disintegration). Mechanical actions cause the breakdown of rock masses, through a combination of grinding, shattering, and breaking, into discrete particles having the same mineral composition as the parent rock. The particles are broken into various sizes and can become angular or rounded. The major agents of physical disintegration are (a) temperature changes causing differential expansion and contraction; (b) freezing of water in pores and cracks; (c) abrasion due to the impact of particles moved by wind, water, air, or gravity; and (d) glaciation, the grinding, plucking, and plowing action of glaciers.

Chemical-biological Weathering. The processes that change the chemical composition of rock include (a) oxidation, the formation of new oxides; (b) hydration, the formation of hydrates; (c) carbonation, in which carbon dioxide from the air forms carbonic acid; and (d) solution, in which organic acides from plants and animals dissolve soluble minerals.

The resulting particles become *clay*, generally complex hydrous aluminum silicates, which are extremely small particles that are formed in sheets or platelets, surrounded by electrically bound water. Clay particles exhibit cohesive attraction to each other because of electrical surface forces. The major clay types are kaolinite, illite, and smectite (montmorillonite).

Weathering Products. Each of the rock types results in a weathering product, *soil*, the character of which depends on the mineral composition of the parent rock and on the weathering process. Typically,

1. Granite becomes tan and yellow silty sand and sandy silt, with some mica and kaolinite clay.
2. Basalt becomes clay, often highly plastic montmorillonite clay.
3. Sandstone and siltstone become sand and silt.
4. Shale becomes silt and clay.
5. Limestone becomes insoluble residue, silicates, and clays.

The weathering of a rock proceeds from the exposed surface downward; thus, the greater the depth, the smaller the amount of weathering. Figure 6.2 illustrates a typical weathering profile in an igneous rock. In a temperate climate, the A and B horizons of the residual soil profile are typically from one to four feet thick. The C horizon, the fully weathered rock (engineering soil) can extend down from a few feet to tens of feet. It grades downward into (a) soil with some rock fragments, then into (b) discrete blocks of rock surrounded by a matrix of soil, and finally (c) into fragmented and jointed rock, ultimately reaching solid rock. Figure 6.3 is a typical weathering profile for metamorphic rocks and Figure 6.4 is a typical weathering profile for carbonate rocks.

The degree of weathering is a function of both temperature and slope. Weathering on gentle slopes in the tropics might be several hundred meters (hundreds of feet) thick. In the far north, the weathered soil zone is very thin. Furthermore, erosion and the plowing action of glaciers have removed most of the weathered soil in places such as the Canadian Shield. Soils formed on steep slopes tend to move downhill under gravity forces and then are eroded by other forces. On shallow slopes, the movement is less severe and the weathered soil zone is much deeper.

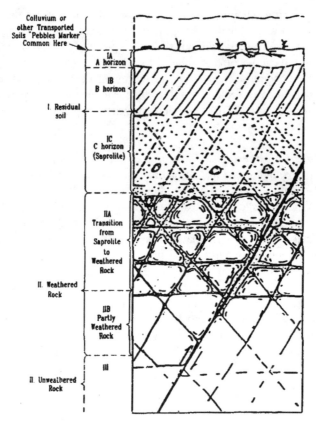

FIGURE 6.2 Typical weathering profile for intrusive igneous rocks [1].

Now, at what point does rock become soil? This is important in contracting for excavation and grading. According to Sowers: "Rock is defined by the engineer as any indurated material that requires drilling and blasting or similar methods of brute force for excavation." [2]

NATURAL SOIL DEPOSITS

Deposits of natural soils can be uniform or non-uniform horizontally and are usually non-uniform, or *stratified*, vertically. The variation in particle size, shape, and texture depends on the weathering process and on the presence or absence of a transportation process. Soil deposits can be *residual* (sedentary) or *transported*.

Residual Soils

Residual soil, or *residuum*, is the fully or partially weathered rock material that has remained in the place in which it was formed. That is, it has not been acted upon by any

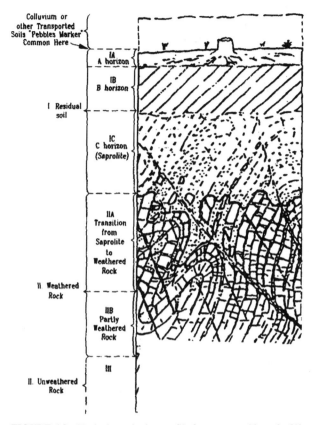

FIGURE 6.3 Typical weathering profile for metamorphic rocks [1].

of the transport mechanisms such as water, wind, gravity, or glaciation. It is found wherever the rate of weathering exceeds the rate of removal. Often, all but the least soluble constituents have been removed by leaching, leaving a clayey material. Because the amount of weathering of a residual soil deposit varies with depth, the index properties (the grain size distribution and the Atterberg limits) also vary with depth. The bouldery rock fragments found in the transition zone often are almost as weathered as the soil. *Saprolite* is a term used to designate a residual soil in which the structure, or fabric, of the unweathered rock is preserved. Figures 6.2, 6.3, and 6.4 are typical residual soil profiles. Deposits of residual soil are generally found on the sides and tops of mountains or high hills.

Transported Soils

Soils are moved from their sedentary, or residual, location by the forces of water, wind, gravity, or ice. The horizontal and vertical variation in particle size, shape, and texture within each type of soil deposit depends on the type and intensity of the process.

When soil particles are moved by either water or air, the *competence* of the fluid is the maximum size of the particle that can be transported at a given fluid velocity. Figure 6.5

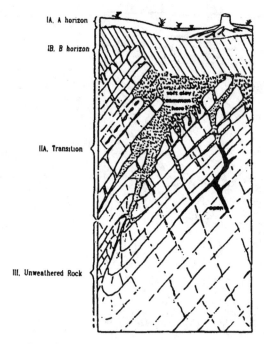

NOTES:
1) Very impure (sandy or silty) carbonates may develope a saprolite, IC zone.
2) A partly weathered chalky limestone, IIB zone is sometimes present.

FIGURE 6.4 Typical weathering profile for carbonate rocks [1].

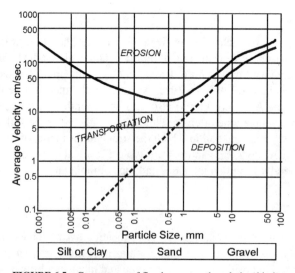

FIGURE 6.5 Competence of flowing water; the relationship between grain size and water velocity needed to cause erosion, transport, or deposition [3].

is a theoretical study of the relationship between water velocity and particle size, showing zones of erosion (suspension in the fluid), transportation (movement along the bottom by jumping or saltation), and deposition (no movement). This diagram also illustrates the vulnerability of various sizes of particles to water or wind erosion.

The coarse sand and gravel particles in the deposition zone are too large and heavy for movement except by very high water velocities, such as those occurring during spring floods in small streams. Clay and silty clay particles have sufficient interparticle attraction due to the clay. Only flocs or clods can be eroded.

Water. Flowing water erodes soils and then deposits them. The tumbling action during saltation produces grain roundness. The competence of a stream of water to move soil particles depends on velocity, grain size, and bed load. At high velocity, such as in flood, particles are eroded and carried in suspension. Then, as the velocity decreases, the particles are redeposited as sediments.

Fluvial (riverine) deposits. Grain size of the bed load along the length of the river channel varies with distance and with changes in the velocity of the stream, which is due to either seasonal variation in water quantity or with curvature of the stream. Large and erratic variations occur transverse to a stream profile, particularly at bends.

Floodplain (alluvial) deposits. Layers of sandy silt, silt, or silty clay are formed on the lowland bordering the stream when the sediment-laden stream overflows its banks. Floodplain deposits tend be uniform in grain size, shape, and texture.

The complex pattern of floodplain deposits shown in Figure 6.6 is typical of the Mississippi and other large, geologically mature rivers. Figure 6.7 is a cross section through part of the deposits of Figure 6.6. *Point bar* deposits are arcuate ridge and swale deposits of sand and gravel formed on the inside of a meander where a slowed velocity causes sedimentation. *Backswamp clays* are the alluvial sediments deposited in the floodplain during a flood and left behind when the water recedes. Natural levees are the sandy silts and silty clays that form flat ridges adjacent to the river. *Clay plugs* are masses of silt, clay, and organic muck filling a cutoff, or oxbow, lake.

Lacustrine (lake) deposits. When a stream enters a large body of water, the velocity slows and the coarsest sediments are deposited in a *delta* that has a uniform grain size. The finer particles are carried farther out, where they settle to form alternating layers of coarse and fine grains.

Coastal (shore) deposits. Sand and gravel are deposited as deltas closest to the point at which a stream empties into the ocean and loses velocity. Waves erode exposed shoreline rock deposits. Longshore currents move the eroded coarse particles with currents, causing separation into various grain-size fractions.

Marine (ocean) deposits These deposits are similar to lacustrine deposits. Sand is deposited next to the shore, silt a little farther out, and clay a great distance from shore, resulting in separation by size.

Uplifted marine terrace deposits. A narrow coastal strip is formed of material deposited in a wave-cut bench in the near off-shore of an ocean and later exposed by tectonic uplift or by lowering of the sea level. Such deposits are typically uniform horizontally but grade down from fine-grained silt at the surface to thick layers of sand and basal gravels. They are generally much less than 50 m (150 ft) thick.

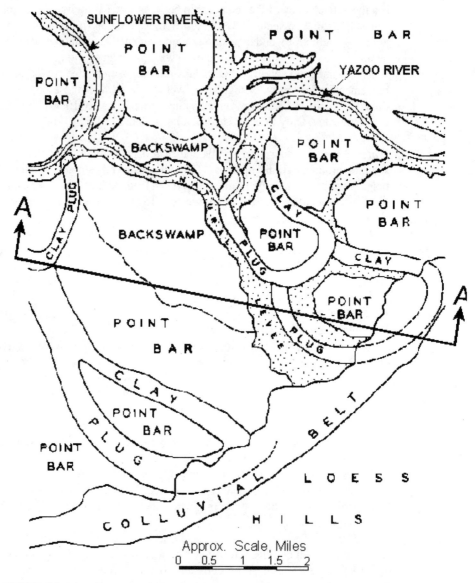

FIGURE 6.6 Meander belt deposits north of Vicksburg, Mississippi, and east of the Mississippi River [4].

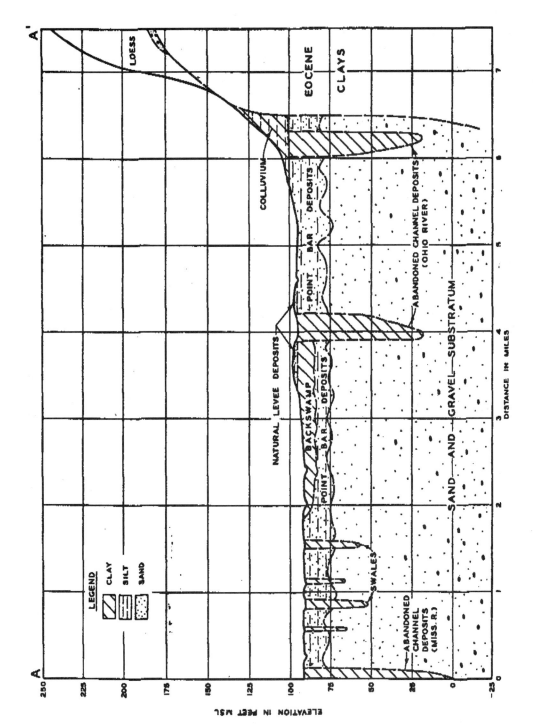

FIGURE 6.7 Section A-A of Figure 6.6. Cross section showing low-lying meander belt deposits and loess-covered valley walls [4].

Wind. Soils in floodplain and coastal deposits are picked up by the wind during dry seasons. The competence of the wind is similar to that of water, shown in Figure 6.5. Downwind, the coarsest particles are deposited closest to the source, often on a bluff line that significantly reduces the wind velocity. The smaller particles are deposited farther away in the direction of prevailing winds.

Sand Dunes. Size of sand grains is fairly uniform and the sand is clean with little or no clay binder. Grains tend to be rounded due to having tumbled in streams. Major dune deposits are found on the southern Oregon coast, on the Indiana shore of Lake Michigan, and in southern Arizona.

Loess Consists of fine sand to silt sizes, with a coating of plastic clay, that were eroded from dry floodplains by the wind (typically from the northwest) and deposited on downwind bluffs and uplands. The competence of the wind diminishes as the wind velocity decreases at a bluff line and farther inland. Loess forms as a metastable or honeycombed structure because of the small clay coating on the silt grains. Loess deposits generally have a low density, but derive their strength from partial saturation. The structure tends to collapse on saturation. Major deposits are found along the south side of the Missouri River and the east side of the Mississippi River. Smaller deposits occur in eastern Washington. The density increases, and the collapse potential decreases, with distance southward down the Missouri-Mississippi valley. Loess deposits have high vertical porosity and high vertical cleavage.

Pyroclastics. Although not true wind-deposited soils, pyroclastics are fragmented rock material and ash formed by volcanic explosion or aerial expulsion from a volcanic vent and deposited nearby. The largest fragments are called *bombs* and the finest are *volcanic ash*. The deposits can be stratified or well-graded. Fresh volcanic ash is a lightweight, porous sand or sandy gravel. The ash can weld together when it is deposited hot, or it can form a soft rock known as *tuff* when it is partially decomposed, wetted, and then dried.

Glaciation. Glaciers grind, pluck, and plow the rocks and soils as they pass. When glaciers melt and retreat, they leave behind deposits of ground rock, clay, and blocks of ice known as *drift*. The material laid down by glaciers is known as *glacial till*, the mixture of the gravel, sand, silt, and clay ground up and carried within the glacier ice. Materials carried by meltwater and deposited downstream are known as *glacial outwash*. Various types of glacial deposits are discussed below.

Ground Moraines. Sometimes known as till plains, ground moraines are the irregular layers of glacial till left in place when glaciers melt. They vary widely in their texture, plasticity, and engineering properties.

Terminal Moraines These are mounds of glacial till that were pushed by glaciers acting as bulldozers and left behind when the glaciers retreated. They can contain small, localized deposits of glacial outwash.

Glacial Outwash This is soil debris, typically sand and small gravel, carried by meltwater and deposited in valleys or broad sheets in front of a terminal moraine. They are sorted by the variable velocity of the flowing water.

Eskers These are sinuous ridges of gravel and coarse sand. Remnants of rivers that flowed in tunnels beneath the ice, eskers are good sources of granular fill materials.

Kames Irregular deposits of gravel and coarse sand formed along edges of ice sheets are called kanes. They are good sources of granular fill materials.

Varved clay is a lake deposit that results from the seasonal melting of glaciers. It forms thin layers of silt, deposited in summer, and clay, deposited in winter. Varved clay is very uniform.

Gravity. Soil masses move under the action of gravity because of loss of shear strength, which is mainly due to increased saturation in the rainy season. Other contributing factors are removal of support at the base of the slope or an increased weight on, or of, the material.

Soil Creep This is the slow downhill movement of soils on moderate-to-steep slopes due to softening (decrease in shear strength) caused by increased moisture content following periods of heavy rainfall. Sliding is incipient, but major movement does not occur. Movement is intermittent and often is measured in centimeters or inches per decade. Soil creep often is associated with residual soil deposits.

Landslides These are rapid movements of soil masses downhill on moderate to steep slopes. Landslides and slope stability will be discussed further in Chapter 12. Landslides result in somewhat heterogeneous mixtures of uphill deposits.

Colluvium This is the accumulation of soil at the foot of a slope and is caused mainly by gravity, often as the result of extensive creeping of residual soils. Talus is included in this broader term. The soil is fairly homogeneous.

Talus Finally, talus is the accumulation of rock fragments, usually angular, and soil at the base of steep rock and/or soil slopes. It occurs because of erosion and/or landslides. It results in a heterogeneous mixture of uphill deposits.

SOIL STRUCTURES

The structure, or fabric, of the components of natural soils depends on the soil minerals, the amount and type of weathering, and on the deposition process. The two major soil type groups, cohesionless and cohesive, react differently to direct compression (see Chapters 4 and 12), to shear deformation (see Chapter 5), and to vibration (see Chapters 7 and 12).

Cohesionless Soil

Coarse grained soils, Figure 6.8, consisting of gravel, sand, and inorganic silt particles, have insignificant amounts of clay binder. Because the weight of individual grains is so much greater than molecular forces, they derive shear strength from interparticle friction and particle interlock. Depending on the mode of deposition and subsequent physical disturbances, they can be either loose or dense. Because direct compression increases shearing resistance, cohesionless soils are best densified by vibration that causes the grains in a loose, unstable structure to move into a denser, more stable structure.

Cohesive Soil

Fine-grained cohesive soils, Figure 6.9, mainly are composed of clay particles, the character of which dominates the shear strength. The interparticle forces of the individual grains due to gravity are insignificant compared to surface molecular forces of attraction and repulsion. They derive strength, and resistance to compression, mainly from the degree of packing (density) and the orientation of the clay platelets relative to each other.

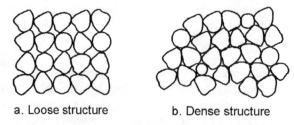

a. Loose structure   b. Dense structure

Figure 6.8 Structures of cohesionless soils (after Sowers [2]).

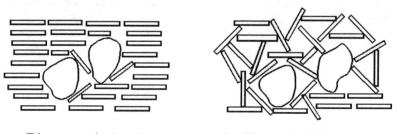

a. Dispersed structure   b. Flocculated structure

FIGURE 6.9 Structures of cohesive soils (after Sowers [2]).

Mixed Grain Soil

Mixtures of coarse and fine grains, Figure 6.10, will have a behavior dependent on the dominant soil type under a shear stress. If there is little clay binder, the mixtures will behave like a cohesionless soil, with shear strength derived mainly from interparticle friction and particle interlock. When the clay content is sufficiently large that the coarse particles are prevented from any significant interparticle friction and particle interlock, then the mixed grain soil behaves like a cohesive soil. The clay content at which one type of behavior dominates over the other often is indistinct and gradational.

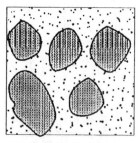 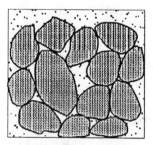

a. Properties of matrix govern behavior.   b. Mechanical interlock governs behavior.

FIGURE 6.10 Structures of mixed grain soils containing coarse and fine grains (after Sowers [2]).

ENGINEERING GEOLOGY OF ROCKS AND SOILS **6.13**

EXAMPLE PROBLEMS

EXAMPLE 6.1

Question:
In what ways does the engineering profile of a typical residual soil deposit formed in igneous rock differ from an alluvial deposit formed adjacent to a river? Discuss variations in index properties and engineering properties horizontally and vertically.

Solution:
Because the amount of weathering of a residual soil deposit varies with depth, the index properties, including the grain-size distribution and the Atterberg limits, also vary with depth. The position of the water table determines the partial saturation zone and affects the shear strength. Shear strength and compressibility also vary within fairly short depths but are fairly uniform over short distances.

Because the pattern of deposition of fine-grained materials on an adjacent floodplain occurs almost annually (if no levees are in place), the stratification tends to be uniform and the soils filling a wide floodplain also tend to be uniform laterally. Shear strength tends to be relatively low and compressibility tends to be high.

EXAMPLE 6.2

Question:
What was the major transportation agent for each of the following sedimentary soils: (a) sand dunes, (b) alluvium, (c) glacial till, (d) volcanic ash, (e) talus, (f) varved clay.

Solution:

a. Sand dunes are wind-eroded beach sands, picked up by the wind and deposited on nearby land.
b. Alluvium consists of the layers of sandy silt, silt, or silty clay that are formed on the lowland when a sediment-laden stream overflows its banks.
c. Glacial till is the mixture of gravel, sand, silt, and clay ground up and carried within glacier ice and deposited when the ice melted.
d. Volcanic ash is the fine-grained ash formed by volcanic explosion or aerial expulsion from a volcanic vent and deposited on nearby surfaces.
e. Talus is the accumulation of rock fragments, usually angular, and soil at the base of steep rock and/or soil slopes.
f. Varved clay consists of lake deposits from seasonal meltings of a glacier that result in alternating thin layers of silt, deposited in summer, and clay, deposited in winter.

EXAMPLE 6.3

Question:
Indicate areas where one would find the following surficial features or soil deposits: (a) terminal moraine, (b) dune sand, (c) loess, (d) alluvium, and (e) colluvium?

Solution:

a. A terminal moraine is found where the movement of a glacier terminated. Therefore, one would be found along the edges of the areas where the glaciers from the four most recent glacial periods reached: namely, along the Canadian border in the northwest, to southern Illinois in the midwest, and in major parts of New England.
b. A sand dune is always found next to an ocean or lake beach. The best known are those along the Oregon and Washington coasts and along the southern end of Lake Michigan.
c. Loess is found on bluffs along the downwind side of alluvial flood plains. The downwind bluff areas of the Missouri River and the Mississippi River contain the most extensive deposits.
d. Alluvium is deposited from overbank flows of flooded rivers. Alluvial deposits are formed along nearly all rivers and can range from a few decameters (a few hundred feet) to nearly 200 km (130 miles) in width.
e. Colluvium generally is associated with slopes in fine-grained soils where soil creep and/or sliding typically occurs. Talus, a form of colluvium, forms as a fan wherever major erosion occurs in rapidly moving streams in hillsides.

EXAMPLE 6.4

Question:
Which has the greatest potential landslide risk: (a) an interior sand dune (that is, one away from an ocean or lake) having two horizontal slopes to one vertical slope, (b) a 5 m (15 ft) high vertical cut in loess, or (c) a cut in a shale deposit, the bedding of which is inclined toward the open space? Explain your answer.

Solution:
The cut in the shale has the greatest landslide risk.

a. A sand dune having a 2:1 slope is generally stable and has a low potential landslide risk. The most likely cause of a landslide in a dune is the removal of sand from the toe, causing the slope angle to increase above the natural angle of repose.
b. A 5 m (15 ft) high vertical cut in loess is generally stable and has a low potential landslide risk. The stability of loess is dependent on shear strength derived from partial saturation. Experience in the Mississippi Valley and other locations has shown that vertical slopes as high as 6 m (20 ft) can be stable for many decades unless the water content is increased materially.
c. The part of a shale deposit exposed in a cut begins to deteriorate and soften due to contact with the air. If the bedding of the deposit is inclined at an angle to the cut, then there is a likelihood that sliding might occur, sooner or later.

REFERENCES

1. Deere, D. U. and Patton, F. D. 1971. "Slope Stability in Residual Soils," *Fourth Panamerican Conference on Soil Mechanics and Foundation Engineering.* American Society of Civil Engineers. New York.

2. Sowers, G. F. 1979. *Introductory Soil Mechanics and Foundations: Geotechnical Engineering.* Fourth edition, p.3. New York: Macmillan Publishing.

3. Hjulstrom, F. 1939. "Transportation of Detritus by Moving Water," in Part 1. Transportation, *Recent Marine Sediments, A Symposium,* Parker D. Trask, Ed. American Association of Petroleum Geologists. Tulsa, Oklahoma.

4. U. S. Army. April, 1981. "Construction of Earth and Rock-fill Dams," Lecture notes, Corps of Engineers Training Course. Geotechnical Laboratory. U.S. Army Engineer Waterways Experiment Station. Vicksburg, Mississippi.

CHAPTER 7
ENGINEERING SUBSURFACE INVESTIGATIONS

The objective of a geotechnical engineering subsurface investigation is to obtain the most complete and accurate estimate possible of the location, extent, and engineering character of the soils and rocks underlying the project site within the limits of time, money, and practicality.

PLAN FOR A SUBSURFACE INVESTIGATION

The typical subsurface investigation consists of the following steps, usually made in the following order:

1. The geotechnical engineer analyzes completely the project location, the current topography of the site, the configuration of the structure, the magnitude and distribution of structural loads, the desired site configuration and grading, and other support requirements for the structure.

2. All available pre-existing information is reviewed. That information could include geologic literature and maps, both published and unpublished, aerial photos, records of previous geotechnical studies in the project area, and personal experiences others have had with soils in the project area. This is sometimes called a literature search or a desk study.

3. Based on this information, an initial hypothesis of the geotechnical subsurface profile is developed, including the types, configuration, and geotechnical character of the soils and/or rocks. If the available information is sufficient for the project, the site investigation is terminated at this point. Otherwise, an estimate is made of the site configuration and the variability of the engineering properties over the site. If the site is known from extensive prior information to be fairly uniform or to vary in a known manner, a site exploration plan is developed. If the site configuration is not well known or if anomalies are suspected, then a geophysical survey may be appropriate.

4. Geophysical surveys are made to establish the overall configuration of the subsurface deposits. The estimated subsurface profile is amended with the new information.

5. If the amended subsurface profile estimate is not sufficient, a physical geotechnical site exploration plan is formulated. The number and location of the test sites will be dictated by site variability. Sampling or testing depth may be reached by borings or dug pits. At each exploration site, specific depths and specific methods are selected for sampling and testing the subsurface materials. Geotechnical engineering field tests are made and/or samples are secured at various depths for laboratory tests. Visual-manual identification tests are made on the soil and/or rock samples in the field.

6. Samples are transported to the laboratory where index properties tests are made to confirm the field identification. Undisturbed samples, if any, are tested for various engineering properties.

7. The resulting data are summarized in an engineering soil/rock profile (cross section) of the site. The validity of the profile depends on the type and extent of the subsurface investigation made, and on the knowledge and interpretive skill of the geotechnical engineer.

8. A geotechnical engineering evaluation is made using appropriate theoretical and empirical behavior models. The engineer evaluates the interaction of the soil and rock deposits within a significant depth in the soil profile with the structural loads. Quantitative recommendations are made for the design of the substructure elements and for the preparation and grading of the site.

9. An evaluation is made of the value of additional field information compared to the cost of obtaining the additional information. Then, if appropriate and if authorized, the engineer returns to step 4 or 5. Otherwise, the field investigation is terminated.

SOURCES OF PRE-EXISTING INFORMATION

Geotechnical site investigations start with an estimate of the soil profile based on the existing information from the geologic literature and maps, former project records, general sources, and, possibly, remote imagery, including aerial photographs.

Geologic Data Sources

Geologic studies include a review of the geologic literature and related records for the project area. Sources of geologic data include those discussed below.

U.S. Geological Survey (USGS). The USGS continually publishes maps, reports, circulars, open-file reports, professional papers, and bulletins covering most of the United States. The Earth Resources Observation System (EROS) provides access to the NASA LANDSAT satellite imagery data.

USDA Natural Resource Conservation Service (NRCS). Formerly called the Soil Conservation Service, the NRCS publishes maps and reports that are intended primarily for agricultural purposes and describe the uppermost five feet of the soil profile. The reports typically contain discussions of near-surface geology.

State Geological Surveys. All states have a state geological survey that publishes maps, reports, and other documents about the geology and mineral resources of that state. The work often is similar in scope to that of the USGS.

Project Records

The documentation for many state and federal projects contains a description of site investigations, analyses of data, construction drawings, and data references, all of which may be useful is establishing the preliminary soil profile at a dredging project site. The General Design Memorandum for each Army Corps of Engineers project contains a summary of the geologic and geotechnical information available for use in the design of that project.

Remote Imaging

Remote imaging, or sensing, is the process of obtaining information about an object using naturally occurring or man-made electromagnetic radiation. Aerial and satellite photography, using either visible or nonvisible light waves, and ground-probing radar are typical of this method.

All materials exhibit an electromagnetic reflection signature and, given sufficient correlation with ground truth tests, can be reasonably identified from remotely acquired images. Light waves cannot penetrate the surface but radar can penetrate to shallow depths if the surface layer is more penetrable than lower layers. Therefore, this method is restricted to the evaluation of surface deposits. It does not permit evaluation of underwater soils, although turbidity plumes, temperature changes, and abnormalities in the character of the water can be defined.

General Sources

In addition to the direct sources of geologic data given above, several general sources contain local information not available elsewhere.

Libraries. Libraries at state and private colleges and universities offering geology, geological engineering, and engineering geology programs have extensive holdings of USGS, SCS, and state geological survey publications. Geology-based theses and dissertations are also on file. Bibliographic indexes of geologic publications and doctoral dissertations from other schools are available, often as part of computer search facilities.

Local and Regional Agencies. Local and regional planning boards sometimes will authorize geologic studies as part of an overall plan for an area. Such studies generally are made by private geologic or geotechnical firms; therefore, the information obtained might contain data and analyses not available from government agency publications.

Knowledgeable Individuals. In the initial stage of a geologic data search, knowledgeable individuals have information about references and a general geologic overview of the project area. People to contact include university professors, reference librarians, geotechnical engineering firms, site exploration firms, local quarry operators, construction aggregate suppliers, and appropriate persons from agencies such as state geological surveys, the USGS, and the U.S. Army Corps of Engineers.

ENGINEERING GEOPHYSICAL METHODS

Geophysical surveys have three objectives: the measurement of geologic features, the in situ determination of certain limited engineering properties, and the detection of hidden cultural features [1]. Geologic features may include faults, bedrock surfaces, discontinuities and voids, and groundwater. Engineering properties that can be determined in situ include elastic moduli, electrical resistivity, and density. Hidden cultural features include buried underground tanks and pipes, contaminant plumes, and landfill boundaries.

Geophysical surveys can provide continuous stratigraphic information about the subsurface over large areas much less expensively than invasive methods such as test pits or borings. The measurements are necessarily averages for the property being measured. Geophysical methods, however, do not directly measure the parameters normally desired by the engineer. Additionally, ground truth field tests in borings must generally be made for correlation of properties.

The surface-based geophysical methods used in geologic subsurface investigations are given in Table 7.1. Of these, the techniques that are the most useful and most commonly used in geotechnical engineering investigations are electrical resistivity, seismic refraction, and ground penetrating radar.

Electrical Resistivity Surveys

The resistivity of earth materials is affected by mineralogy, porosity, degree of saturation, moisture content, and the chemistry of pore fluids. Electrical resistivity methods use the differences in electrical resistivities to define subsurface layering and locate cavities, gravel pockets, and the groundwater table. Figure 7.1 shows a typical surface array of four electrodes, two current electrodes, and two voltage measuring electrodes. The average of the apparent resistivities of all the earth materials through which the current is flowing is calculated by Ohm's law. As the electrode spacing is increased, the current flows through more materials and the apparent resistivity calculated from the array is used to characterize the subsurface layering in a manner similar to that used in seismic refraction surveys.

Seismic Refraction Surveys

An energy source, such as a hammer impact on a steel plate or an explosive charge in a shallow hole, is applied to the ground surface. A linear array of geophones picks up the resulting compressional wave and a seismograph determines the wave velocities. Figure 7.2 illustrates the way in which the resulting velocity measurements are used to configure the subsurface stratification.

TABLE 7.1 Surface-Based Geophysical Methods (modified from TRB [2])

Type of Survey	Applications	Limitations
Electrical resistivity	Locates boundaries between clean granular and clay strata, groundwater table, and soil-rock interface.	Difficult to interpret and subject to correctness of the hypothesized subsurface conditions
Electromagnetic profiling	Locates boundaries between clean granular and clay strata, groundwater table, and rock-conductivity mass quality; offers even more rapid reconnaissance than electrical resistivity.	Difficult to interpret and subject to correctness of the hypothesized subsurface conditions; does not provide engineering strength properties.
Seismic refraction profiling	Determines depths to strata and their characteristic seismic velocities.	May be unreliable unless strata are thicker than a minimum thickness, velocities increase with depth, and boundaries are regular. Information is indirect and represents average values.
Direct seismic (uphole, downhole, and crosshole surveys)	Obtains velocities for particular strata, their dynamic properties, and rock-mass quality.	Data are indirect and represent averages; may be affected by mass characteristics.
Microgravity	Extremely precise; locates small volumes of low-density materials utilizing very sensitive instruments.	Use of expensive and sensitive instruments may be impractical, especially in rugged terrain typical of many landslides and remote areas; requires precise leveling and correction for local topographic features; requires detailed information on topography and material variations.
Ground penetrating radar	Provides a subsurface profile; locates buried objects (such as utility lines), boulders, and soil-rock interface.	Uses high frequency (80 to 1000 MHz) EM pulses transmitted from an antenna held close to the ground surface. Pulses are reflected from various interfaces within the ground and detected by the receiver. Depth of penetration ranges from 1 to 30 meters (3 to 100 ft) depending on soil and rock conditions. Has very limited penetration in clay materials.

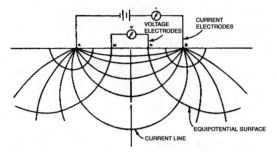

FIGURE 7.1 Arrangement of electrodes to measure resistivity of portion of earth's surface as part of resistivity survey.

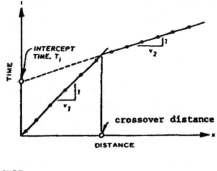

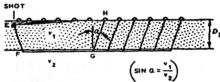

FIGURE 7.2 Simple two-layer case of seismic refraction survey, with corresponding time-distance curve (after Redpath, 1973 [3]).

SUBSURFACE ACCESS METHODS

Methods and devices for making geotechnical subsurface investigations are varied and each has its advantages and limitations. The equipment and technique needed to access sampling or testing depth must be adapted to the expected soil and rock types. Some of the more common subsurface access methods are discussed below.

Pits and Trenches

Pits and trenches can be dug by hand or by a machine, such as a backhoe. The depth is limited by the reach of the backhoe arm to about 15 ft or less. This method permits examination of all strata within the pit above groundwater level. Disturbed (bag) samples, hand-carved undisturbed samples, and manually operated tests can be made. Excavations are limited by caving soils and by groundwater.

Borings

Borings are drilled holes ranging in size from a few inches in diameter to borings large enough for downhole access by a person. Borings permit securing disturbed or undisturbed (tube) samples and/or performance of field tests at any depth within the bore hole. The preferred drilling method depends on the boring depth, equipment capability, presence of groundwater coupled with highly permeable strata, and the presence of very large particles or rock. The usual drilling methods generally cannot penetrate strata containing large gravel or boulders. The more common drilling methods are discussed here.

Hand Auger Hand augers, shown in Figure 7.3, are operated manually. Hand-auger holes range from two to more than six inches in diameter. Continuous disturbed sampling is possible. The hole depths are limited to about 20 to 25 feet.

Bucket Auger. Bucket augers, shown in Figure 7.4, are machine operated. Bucket auger holes range from a few inches to more than five feet in diameter. Buckets are attached to the end of a drill rod and removed from the hole to empty. Continuous disturbed sampling is possible. Bucket augering depth is limited only by equipment capacity and soil or rock types, and typically is about 100 feet.

Continuous Flight Auger. A spiral auger with continuous sections, shown in Figure 7.5, is rotated into the ground, causing cuttings to come to the surface. Augers are typically three to six inches in diameter. The drilling depth is limited by the available drill engine power, but is usually about 75 to 100 feet. Augers must be withdrawn to obtain samples or to make field tests.

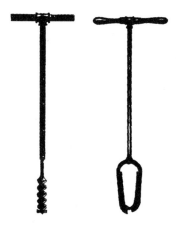

FIGURE 7.3 Spiral-type and Iwan-type hand augers.

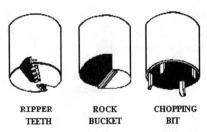

FIGURE 7.4 Examples of bucket augers for various soil types.

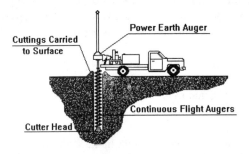

FIGURE 7.5 Continuous flight spiral auger operated by machine.

Hollow Stem Auger. A continuous flight auger is formed around a pipe casing. This has all advantages of an auger (no drilling fluid needed) and of a cased hole and permits sampling or testing without withdrawal of the auger from the hole. The drill stem for the sampling or testing device is inserted and retrieved through the hollow casing. Interior casing sizes range from three inches to more than six inches and exterior sizes correspond. These augers require larger torque than plain flight augers because of their larger size. Drilling depth is limited by the power available, but is usually from 75 to 100 feet.

Wash Boring. The wash boring method, shown in Figure 7.6, uses a chopping bit to cut the soil and a pump to circulate drilling water. The cuttings are returned to the surface by the circulating water. Some versions of the equipment are hand-transportable. The drill stem can be worked by hand and must be withdrawn to obtain samples or make field tests. Drilling depth is limited by pump power available, but is practically about 50 feet.

Rotary (Mud) Drilling. The rotary drilling method, shown in Figure 7.7, is machine operated and uses a rotating bit to cut soils and soft rock. Cuttings are returned to the surface by pumping circulating water containing heavy drilling mud, which is used to keep the hole open against lateral soil or fluid pressure. A driven pipe casing is sometimes used to maintain the opening when drilling fluid is not sufficient. The drill stem, handled by a power winch, must be withdrawn to obtain samples or make field tests. Drilling depth is limited by the pump power available, but is usually from about 100 to 200 feet. Groundwater level cannot be determined in the drill hole unless a special drilling mud, Revert, is used.

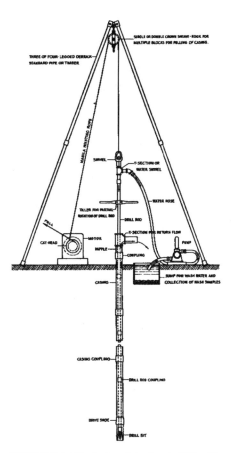

FIGURE 7.6 Typical wash boring rig ([6]; with permission of ASCE [4]).

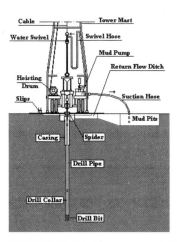

FIGURE 7.7 Typical rotary (mud) drilling rig ([6]; with permission of ASCE [4]).

Air Rotary. Air rotary drills are machine operated. They use a rotating bit to cut soils and soft rock and cuttings are returned to the surface by circulating high velocity air provided by an air compressor. Driven pipe casing can be used to keep the hole open when there is a tendency for it to collapse. The drill stem, handled by power winch, must be withdrawn to obtain samples or make field tests. Drilling depth is limited by available air compressor capacity, but is practically from about 100 to 200 feet.

Becker Drilling. The Becker drill is a patented system that can be used when large gravel or boulders are encountered in the subsurface. The Becker drill breaks large particles with percussion drilling and then returns the smaller particles to the surface using circulating air under high pressure.

SAMPLING OF SOILS AND ROCKS

When the desired depth is reached, by either a boring or a pit, a sample of the soil or rock can be obtained for laboratory tests or field tests. In a *representative* but *disturbed* sample, the *in situ* particle structure is altered by the remolding action of the sampling device; however, the sample contains all of the soil components in their *in situ* amounts. The structure of an *undisturbed* sample is relatively unaltered. Commonly used soil and rock samplers can be grouped as follows:

1. Representative (Disturbed, Remolded) Samplers
    a. Thick-wall split-barrel drive samplers
    b. Thick-wall vibrating tube samplers
2. Undisturbed samplers
    a. Thin-wall tube samplers
    b. Diamond-tipped core-barrel samplers

Thick-Wall Split-Barrel Drive Samplers

Impact, or percussion, is used to drive a thick-wall split-barrel sampler, as seen in Figure 7.8. The commonly used sampler sizes are shown in Table 7.2. The best known of these devices is the split tube (split barrel) sampler used in the standard penetration test, seen in Figure 7.8 [4] [5], which is the 5.1 cm (2.0 in.) OD device of Table 7.2.

TABLE 7.2 Thick-Wall, Split Barrel Samplers

Outside Diameter	5.1 cm (2.0 in.)	6.4 cm (2.5 in.)	7.6 cm (3.0 in.)	8.9 cm (3.5 in)
Inside Diameter	3.8 cm (1.5 in.)	5.1 cm (2.0 in.)	6.4 cm (2.5 in.)	7.6 cm (3.0 in.)
Drive Shoe	All samplers are typically fitted with a hardened steel drive shoe having the same OD as the sampler, but with an inside diameter 0.32 cm (0.125 in.) smaller than the sampler ID. This permits the use of a thin metal sample liner inside the sampling barrel, if desired.			
Length	All samplers normally are 61 cm (24 in.) long, but longer versions are available			

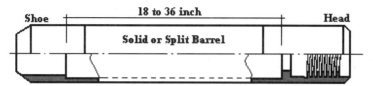

FIGURE 7.8 Thick-wall split-barrel sampler used in the Standard Penetration Test.

Advantages. These devices are capable of penetrating and retaining a variety of soil types and consistencies, and are usually deployed in a small-diameter drilled hole. The maximum size of particle that can be retained is slightly smaller than the inside diameter of the drive shoe.

Limitations. Because of the thickness of the sampler walls, soils squeezed into the tube are disturbed by remolding. These devices require a heavy drop weight and a machine with a mast, shown in Figures 7.5 and 7.7, or a tripod and hand-operated pulley, shown in Figure 7.6, to raise and drop the hammer.

Vibrating Tube Sampler

Another means of inserting a sample tube into a soil deposit is by high-frequency vibration of the sampler during pushing. The vibrating tube sampler uses a long thick-walled tube that is vibrated into the soil by means of a vibrating motor attached to the top of the tube.

There are several manufacturers of vibro-corer devices worldwide. One proprietary device uses high frequency (7000 to 12,000 vibrations per minute) and low amplitude vibrations applied to the drill string to shear the soils in the immediate vicinity of the cutting edge of the core barrel. This permits the device to enter unconsolidated granular and cohesive deposits at rates of up to 1.5 m (5 ft) per minute. The proprietary vibro-corer is lightweight, having a 39 kg (85 lb) engine, an 11 kg (25 lb) drive head, and lightweight tubes of 85 mm and 135 mm (3.35 and 5.31 inches) in diameter.

Advantages. The vibrating tube sampler obtains representative samples in long lengths. The unit is portable and can be operated by a two-person crew from a floating or fixed platform. It does not require a lifting mast or heavy weight.

Limitations. The thick wall causes remolding disturbance of cohesive samples. For sands, the thick wall coupled with the vibration practically assures sample disturbance.

Thin-Wall Tube Samplers

Laboratory strength tests of clays require a true undisturbed sample. A schematic of thin-wall tube sampling process in cohesive soils is shown in Figure 7.9, in which (a) the sampling is started by inserting the sampler into the bottom of the hole, (b) the sampler is slowly pushed (not driven) into the soil, and (c) the sampler with soil is withdrawn.

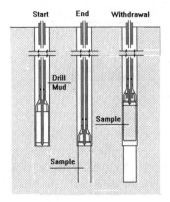

FIGURE 7.9 Sampling using a thin-wall tube sampler.

The requirements for an undisturbed soil sampler for geotechnical testing are given in the classic 1949 report by Hvorslev [6]. Hvorslev graphically showed the effects of sample disturbance and validated the need for thin-wall tubing, with a small area ratio, well-designed inside and outside clearance ratios, and a friction-free interior. Shelby (the trade name of U.S. Steel seamless steel tubing) tubes meeting the Hvorslev criteria are prepared for soil sampling by having a sharpened, crimped end so that

$$\text{Inside clearance ratio, \%} = 100 \frac{D_i - D_e}{D_e} \tag{7.1}$$

is less than about 1 percent, and

$$\text{Area ratio, \%} = \frac{D_o^2 - D_i^2}{D_i^2} \tag{7.2}$$

should be less than 10 percent,

where D_i = Inside diameter of the sampling tube
 D_e = Diameter of the sampler cutting tip
 D_o = Outside diameter of the sampling tube.

Advantages. If a true undisturbed sample is obtained and returned to the laboratory, then the unit weight and water content can be determined and the sample can be tested for shear strength, compressibility, and/or permeability.

Limitations. This method requires a reaction against which it can push. It usually is used with machine drills, shown in Figures 7.5 and 7.7. The use of thin-wall tubes is limited to the undisturbed sampling of soft to stiff clayey materials. Very soft clayey soils will not be retained in the sampler unless extra measures, such as using a piston inside the sampler, are used.

Pushing thin-walled tubes into hard cohesive materials or into gravels or sandy gravels is virtually impossible without seriously crimping the edge or collapsing the tube. Although thin-wall tube samples can be obtained from clean sand, even the small area ratio of such tubes is sufficient to change the relative density and the structure of the sand so that the sample is considered disturbed, or remolded.

Finally, the sample tube must be transported carefully to the laboratory. Once there, the sample must be extruded from the tube. In both instances, there is a high potential for sample disturbance.

Diamond-Core Barrel Sampler

Extremely hard materials, such as shale, cemented soils, and rock, are too hard for sampling by the direct insertion of a thin metal tube. Therefore, an undisturbed core is obtained by fitting the circular end of the sampling tube with a hardened steel cutting surface, or bit. For cutting rock, industrial diamonds are imbedded in the cutting edge of the bit. Hydraulic pressure and rotation cause abrasion of the rocky material from the annular space between the core and the wall of the drill hole. Water or drilling fluid is circulated down the drill stem, between the core and the inner face of the *single-tube core barrel*, and then back up the hole to cool the bit and to return the cuttings to the surface. The core is retained in the core barrel and retrieved. This device is not often used commercially because of the greater adapability of the double-tube core barrel.

When the rock is erodible, because of softness due to decomposition or of interlaminated soil materials, a *double-tube core barrel*, shown in Figure 7.10, must be used. In this system, two concentric tubes are used. The inner core barrel does not rotate; the drilling fluid flows between the inner and the outer, diamond-tipped, barrel. The inner barrel protects the rock core from the eroding water. This is the most commonly used rock-sampling device. For highly fractured rock, a *triple-tube core barrel* is used. A third, concentric, inner, longitudinally split tube, facilitates removal of the sample from the tube.

Soft rock, cemented soils, shale, and hard clays can be drilled with a hardened steel serrated bit instead of diamonds in the tip of the core barrel. The *Denison sampler*, seen in Figure 7.11, is similar to a double-tube core barrel except that the inner, non-rotating tube projects beyond the outer, rotating tube so that the drill fluid will not come in contact with, and erode, the material being sampled. The amount of projection can be adjusted for the type of material being sampled. The *Pitcher sampler* differs from the Denison sampler only in that the inner tube is spring controlled.

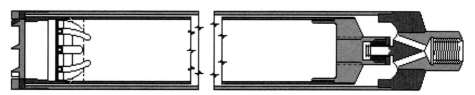

FIGURE 7.10 Double-tube core barrel for sampling rock.

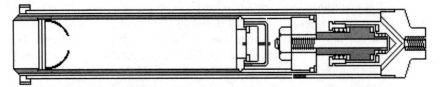

FIGURE 7.11 Denison sampler for hard clays and soft rock.

IN SITU TESTING OF SOILS AND ROCKS

Two types of tests are used to estimate the in situ shear strength of a soil.

1. Direct tests, which measure the shear strength by direct simulation of field shearing conditions, are:
   a. Plate load test
   b. Field vane shear test (VST) of cohesive soil
   c. Borehole shear test (BST)
   d. Unconfined compression test of undisturbed cohesive sample
   e. Handheld devices: Penetrometer and/or Torvane test of cohesive sample
2. Indirect, empirical tests used with empirical correlations to estimate shear strength are:
   a. Standard penetration test (SPT)
   b. Dynamic penetrometer test, thick-wall tube or solid cone
   c. Static-cone penetration test (CPT)
   d. Hand-held sounding rod test

Plate Load Test

The plate load test shown in Figure 7.12 is the original engineering field strength test for soils. It typically is made on a small (1 ft^2) steel plate, usually at foundation level. The test requires heavy reaction, normally a dead load of some sort. The test load is applied in increments to failure or to a maximum load or deformation in a manner similar to the pile load test. The test can be made in a dug pit or at bottom of large diameter drilled shaft. A larger plate load test sometimes is used to determine the coefficient of subgrade reaction for evaluating pavements and beams on elastic foundations (see also the discussion of the modulus of subgrade reaction in Chapter 8).

Advantages. The zone of soil tested by a small test plate is generally within one to one and a half times the width of the footing. If the soil stratum being evaluated is relatively uniform in shear strength, or varies in a known manner, and if the test is made on truly undisturbed material, then the test indicates the actual shear strength and can be evaluated in terms of theoretical bearing capacity equations.

Limitations. As shown in Figure 7.13, however, if a softer, more compressible material exists below the zone of influence of the test plate, the plate load test may not be affected

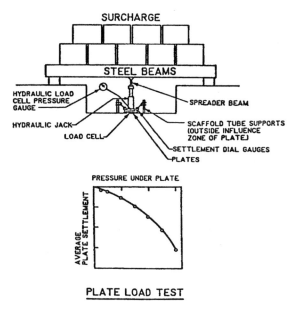

FIGURE 7.12 Plate load test on a small test plate [7].

by its presence, yet the larger actual foundation may be seriously affected. Because they are made relatively quickly, plate load tests in fine-grained soils do not reflect the long-term displacement due to compression settlement.

Except in the rare case of a massive, crack-free material, load tests on rock are rarely of value. The load-carrying capacity of rock is controlled by the degree of jointing, weathering, bedding, and cementation in the rock. These effects are so variable that a load test on a small plate does not permit meaningful extrapolation of behavior to the larger bearing area under actual footings.

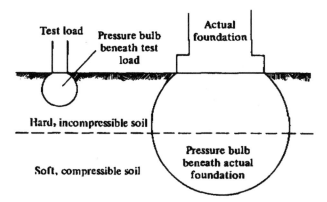

FIGURE 7.13 Limitations of plate load test. The loaded test plate is not affected by the softer layer, although the full-size footing is affected [7].

Field Vane Shear Test

The device for making a field vane shear test consists of four thin blades, two to four inches in diameter and attached to the end of a drill rod, that are inserted into the undisturbed soil below the bottom of a boring and then rotated, as shown in Figure 7.14. The effect is to duplicate a consolidated-undrained (R-test) rotational direct shear test in which the normal force on the shear plane is the *in situ* lateral earth pressure and the torque is the shear force.

Advantages. The VST is used on clayey soils in their undisturbed condition. It does not require removal of a tube sample. In soft to firm clays, the vane can be pushed by hand and the torque applied with a standard automotive torque wrench.

Limitations. The test is only valid for cohesive soils. Because of the thinness of the blades, the practical upper limit of shear strength for the VST is on the order of 200 kPa (2 tsf), or a stiff clay. Interpretation of VST results appears, at first, to be simple and straightforward: a direct shear test has been made *in situ* and the measured shear strength is the undisturbed cohesion. The correction factor of Equation 7.3, however, must be applied to the indicated shear strength to account for clay plasticity.

$$C = 1.7 - 0.54 \, (PI) \tag{7.3}$$

where C = Correction factor
PI = Plasticity index of the soil.

Borehole Shear Test

The borehole shear test, shown in Figure 7.15, simulates the laboratory direct shear test on the walls of a boring. A normal force is applied hydraulically to two shear plates bearing

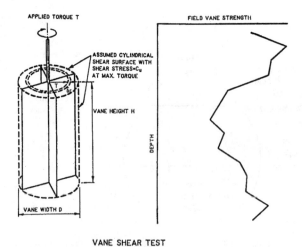

FIGURE 7.14 Field vane shear test device [7].

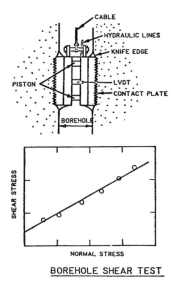

FIGURE 7.15 Borehole shear test device [7].

on opposite sides of a bore hole; a shearing force is applied to the soil on the sides of the hole by a direct pull on the devices. The intent of the test is to measure the angle of internal friction directly on undisturbed soil. Like the laboratory direct shear test, drainage and volume change on the shear plane during shear is a function of soil permeability.

Advantages. In free-draining granular soils, the BST measures the drained shear strength; in cohesive soils, the BST measures the undrained strength, which is equivalent to the unconfined compressive strength. The test does not require a heavy drop weight or a heavy push reaction.

Limitations. The time required for a test sequence of three to four tests at different normal pressures ranges from 20 minutes to two hours, and averages about one hour. Great care must be used in preparing the boring and the equipment for the field test. The test procedure requires a smooth hole of constant diameter in the test area, which is difficult to achieve with common drilling methods.

Unconfined Compression Test of Undisturbed Cohesive Sample

The simplest and most straightforward undrained shear strength test of cohesive soils is the unconfined compressive strength test. This is equivalent to the unconsolidated-undrained direct shear or triaxial shear test discussed in Chapter 5. A cylindrical undisturbed sample from a thin-wall sampling tube, with height twice the diameter, is tested in the device shown in Figure 7.16 in simple compression, without confining pressure, to failure within one to two minutes. Standard terms relating relative consistency and the unconfined compressive strength of cohesive soils are given in Table 7.3.

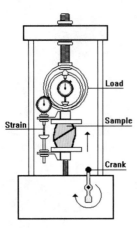

FIGURE 7.16 Laboratory-controlled strain device for unconfined compression tests of soils.

TABLE 7.3 Consistency of Cohesive Soils

Consistency Term	Unconfined Compressive Strength	
	Tons/sq ft	kPa
Very Soft	< 0.25	< 25
Soft	0.25-0.50	25-50
Medium (Firm)	0.50-1.00	50-100
Stiff	1.00-2.00	100-200
Very Stiff	2.00-4.00	200-400
Hard	4.00	> 400

Advantages. This is a direct measure of shear strength. The water content and bulk density of the test sample usually are measured in conjunction with the test.

Limitations. This test requires a true undisturbed sample. The unconfined compressive strength of clays and shales tested at very rapid strain rates, such as those occurring during very rapid slope cutting, increases by 30 percent to 40 percent or more over the strength from the more common laboratory tests made at a slower rate.

Handheld Devices: Penetrometer and/or Torvane Test of Cohesive Sample

Handheld devices, shown in Figure 7.17, often are used to indicate the relative consistency of SPT samples or of chunk samples from backhoe pits. The hand, or pocket, simulates a small footing test by inserting a $\frac{1}{4}$-inch diameter, spring-reaction piston into the sample. By calibration with the bearing capacity of shallow footings in clay, the device indicates the unconfined compression test value, which is twice the unconsolidated undrained cohesion.

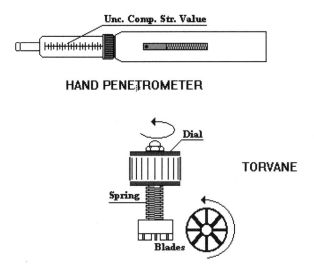

FIGURE 7.17 Handheld devices for estimating the unconfined compressive strength of saturated clays.

The Torvane device is a handheld vane shear device that uses the torque to rotate several small vanes inserted into the surface of a clay sample to indicate the shear strength unconsolidated undrained cohesion.

Advantages. These devices allow the engineer to place a numerical value on an otherwise subjective evaluation of the consistency of a clayey sample in the field, a meaningful adjunct to other identification tests. These tests are equivalent to a *calibrated thumb*. Comparison of the two values permits field or laboratory visual-manual evaluation of clay plasticity.

Limitations. Although both devices presumably yield the cohesion value, their indications will coincide only if the soil is a medium- to high-plasticity clay. The silt and fine sand content of many low-plasticity clays causes the two devices to differ because of the basic theoretical mechanisms involved.

Standard Penetration Test

The SPT is the most widely used field test method for soils. The *in situ* strength and compressibility of soils typically are estimated by means of empirical correlations with the SPT, ASTM D1556. The test involves driving a 2-inch OD, 1.5-inch ID split-barrel sampler into the bottom of a bore hole using a 140-lb. hammer falling freely 30 inches. The number of hammer blows required to drive the sampler for each of three six-inch increments is recorded in the field. The penetration resistance *N-value* is designated as the number of hammer blows required to drive the sampler the final two 6-inch increments.

Recent improvements in our understanding of the SPT have led to corrections to the field-derived SPT values to account for the velocity energy ratio of the hammer release system, usually taken at 60 percent of theoretical free-fall energy; the dynamic efficiency of the hammer; the effect of the rod length; the effect of the sampler type; the effect of the bore hole diameter; and the effect of overburden pressure, $\overline{\sigma}_v$, usually taken as 100 kPa (1 tsf or 2 ksf). The corrected N-value is calculated as

$$(N_1)_{60} = N \times C_e \times C_l \times C_s \times C_d \times C_N \tag{7.4}$$

where $(N_1)_{60}$ = Normalized SPT blow count, for 60% rod-energy ratio and $\overline{\sigma} = 100$ kPa (1 kg/cm²; 1 tsf, 2 ksf)
N = Field SPT blow count, from 6 to 18 inches
C_e = Correction for hammer release system energy (see Table 7.4)
C_l = Correction for rod length (see Table 7.4)
C_s = Correction for sampler type (see Table 7.4)
C_d = Correction for bore hole diameter (see Table 7.4)
C_N = Correction for effective overburden pressure, $\overline{\sigma}$ (Equation 7.5)

$$C_N = (P / \overline{\sigma}_v)^{1/2} \tag{7.5}$$

where P = 100 kPa or 2.0 ksf or 1 tsf, or 1 kg/cm²:

Dynamic Penetrometer Test, Thick-Wall Tube or Solid Cone

Where successive layers vary widely in strength or hardness, the driving of a metal probing device can be used to define relative strength and stratum changes with fair to good accuracy. A cone-tipped penetrometer rod or similar device can be continuously impact-driven using a machine- or hand-operated drop weight. Continuous driving obviates the need to withdraw the rods after each test. Because no sample is obtained, this test method is particularly effective for low-cost, rapid investigation of a sediment where the sediment type and stratification are reasonably well known in advance of testing, from prior experience or geophysical survey. This method may be useful and cost effective in investigating the depth to rock or the very hard layer beneath soft layers or the depth of scour at bridge sites.

Static-Cone Penetration Test

A rod, having an enlarged cone-shaped tip of 1.4 inches diameter, is pushed into the ground at the rate of 2 to 4 feet per minute, as shown in Figure 7.18. The tip's resistance to penetration, Q_t, is measured by means of an electric force transducer attached to the tip. In a friction-cone device, a second pipe rod is placed inside the outer pipe, surrounding the inner rod, with an enlarged friction sleeve behind the cone tip. A separate electric force transducer measures the friction resistance, F_s, of the sleeve. A typical force reaction is a 20-ton truck The soundings and recordings for push forces are continuous.

The friction ratio, calculated as the ratio of the friction resistance to the cone-tip resistance, has been correlated with soil type and is used to estimate the character and description of the soils encountered. An empirical relationship between normalized cone resistance, normalized friction ratio, and soil identification [9] is shown in Figure 7.19.

TABLE 7.4 Corrections to Standard Penetration Test N-values (After Skempton [8])

C_e = Correction for Hammer Release System Energy			
Release Type	**Cathead**	**Hammer**	
(USA) Trip	None	Automatic	1.38
(USA) Slip rope, 2 turns	Large	Safety	0.92
(USA) Slip rope, 2 turns	Large	Donut	0.75
(Japan) Tombi	None	Automatic	1.30
(Japan) Slip rope, 2 turns	Small	Donut	0.90
(UK) Trip	None	Pilcon	1.00
(UK) Slip rope, 1 turn	Small	Old standard	1.00
(China) Trip	None	Automatic	1.00
(China)	—	Donut	0.83

C_l = Correction for Rod Length					
Drill rod length, meters	=	3-4	4-6	6-10	over 10
Drill rod length, feet	=	10-13	13-20	20-33	over 33
Rod length correction, C_l	=	0.75	0.85	0.95	1.00

C_s = Correction for Sampler Type				
Split Barrel Sampler Type	=	Without Liner (1½" ID)	With Liner (1-⅜" ID) Dense sand, clay	With Liner (1-⅜" ID) Loose sand
Sampler Correction, C_s	=	1.0	0.80	0.90

C_d = Correction for Bore Hole Diameter*				
Bore Hole Diameter, cm	=	6.5-12	15	20
Bore Hole Diameter, in.	=	2.5-5	6	8.25
Bore Hole Correction, C_d	=	1.0	1.05	1.15

* C_d = 1.0 for all diameters of hollow stem augers where SPT is taken through the stem.

Olsen [9] [10] considers that the variable stress exponent is the most advanced stress normalization technique for shallow and deep situations. The use of the variable stress exponent requires iteration and is most efficiently accomplished with a computer program. Normalized cone and sleeve friction resistances are

$$q_{c1e} = \frac{q_c}{(\sigma'_v)^c} \tag{7.6}$$

$$f_{c1e} = \frac{f_s}{(\sigma'_v)^s} \tag{7.7}$$

$$R_f = 100\left(\frac{f_s}{q_c}\right) \tag{7.8}$$

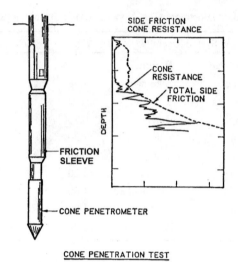

FIGURE 7.18 Static-cone penetration test device [7].

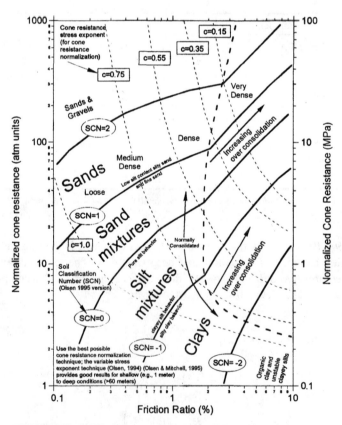

FIGURE 7.19 CPT soil characterization chart [Olsen, 10].

where σ'_v = Vertical effective stress (1 atm, 1 tsf, or 100 kPa)
q_{c1e} = Normalized cone resistance
q_c = Measured cone resistance (1 atm, 1 tsf, or 100 kPa)
c = Cone resistance stress exponent
f_{s1e} = Normalized sleeve friction
f_s = Measured sleeve friction (1 atm, 1 tsf, or 100 kPa)
R_f = Friction ratio, percent.

The vertical effective stress for normalization is taken at 1 atm, 1 tsf, or 100 kPa. The CPT stress exponent is determined from Figure 7.19 by iteration. For linear hand-calculation for depths between 4 and 10 meters (13 and 33 feet), the stress exponent can be roughly estimated as 0.6 for sand and 1.0 for clay. The suggested [11] technique for making an iterative solution of the normalized CPT parameters is outlined as follows:

1. Make an initial estimate of q_{c1e}, from Equation 7.6, guessing if necessary. For a computer solution, use 0.56.
2. The calculated values of q_{c1e} and R_f are then plotted on the CPT soil classification chart in Figure 7.19.
3. Note the corresponding value of the stress exponent c. In most cases, this c will not equal the previously used value. For calculation stability, the next value of c should be between the previously assumed value and the value obtained from the plotted point.
4. Calculate a new estimated value of q_{c1e}, using the c value from Step 3.
5. Return to Step 2 and carry out the next iteration.

Generally three to six iterations are required before the estimated value from the calculation comes within 5 percent of the chart value. The resulting point will not only indicate the soil type but whether the soil is normally consolidated or overconsolidated.

Handheld Sounding Rod Test

Where successive strata vary widely in strength or hardness, the pushing or driving of a simple probing device, such as a rod or steel reinforcing bar, can be used to define the stratum changes with fairly good accuracy. This test method is particularly effective for a low-cost, rapid investigation of the surface of a hard layer or rock. No sample is obtained.

ANALYSIS OF PENETRATION TEST DATA

Several estimates of the engineering properties of soils are made possible by the analysis of the results of SPTs and CPTs made in the field. Correlations exist for estimating the unconfined compressive strength of cohesive soils and the relative density and drained friction angle for sands.

Estimating Unconfined Compressive Strength from SPT Data

Figure 7.20 shows the use of the SPT to estimate the unconfined compressive strength of a cohesive soil. This relationship has considerable scatter in the initial data and should be used with caution.

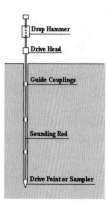

FIGURE 7.20 Handheld sounding rod [6].

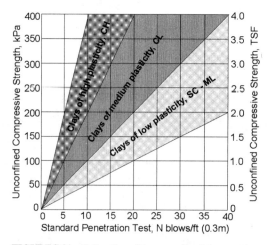

FIGURE 7.21 Estimation of the unconfined compressive strength of saturated clays from the SPT and the Atterberg limits

Estimating Relative Density and Friction Angle from SPT Data

The relationship of the SPT value to relative density of sands is given in Table 7.5, based on work by Skempton [8]. Various correlations have been proposed for the relationship between the SPT and drained friction angle ϕ'. The relationships shown in Table 7.6 [9] are based on 1978 work by Schmertmann [13]. Terzaghi, Peck, and Mesri [4] presented empirical relationships that can be reasonably approximated by a straight line for N-values up to 50 blows per foot (0.3 m):

$$\text{For coarse-grained sands:} \quad \phi' = 30° + N/3 \tag{7.9}$$

$$\text{For fine-grained sands:} \quad \phi' = 28° + N/4 \tag{7.10}$$

These relationships underestimate ϕ' for calcareous sands with crushable particles and overestimate ϕ' for overconsolidated sands.

TABLE 7.5 Relative Density of Sands Based on Standard Penetration Test [8]

Term	Relative Density, percent	Normalized* SPT –values	
		Natural Deposits[†]	Recent Fills[†]
Very loose	0-15	0-3	0-2
Loose	15-35	3-8	2-6
Medium (firm)	35-65	8-25	6-18
Dense	65-85	25-42	18-31
Very dense	85-100	42-58	31-42

* Corrected to 60% of free-fall energy of standard hammer weight and drop and normalized to unit effective overburden pressure of 100 kPa (1 tsf, 2 ksf).
† 1. Natural deposits have been in place (undisturbed) for more than 100 years.
  2. Recent fills have been in place for about 10 years.

TABLE 7.6 Angle of Internal Friction of Sands [12].

(a) Relative Density and Gradation

Relative Density D_r, %	Fine-Grained		Medium-Grained		Coarse-Grained	
	Uniform	Well-Graded	Uniform	Well-Graded	Uniform	Well-Graded
40	34	36	36	38	38	41
60	36	38	38	41	41	43
80	39	41	41	43	43	44
100	42	43	43	44	44	46

(b) Relative Density and *In Situ* Soil Tests

Soil Type	Relative Density D_r, %	Standard Penetration Resistance N_{60} (Terzaghi and Peck [14])	Cone Penetration Resistance q_c, ksf (Meyerhof, [15])	Friction Angle, ϕ' degrees		
				Meyerhof, [15]	Peck, Hanson, & Thornburn [16]	Meyerhof, [17]
Very Loose	< 20	< 4	—	< 30	< 29	< 30
Loose	20-40	4-10	0-100	30-35	29-30	30-35
Medium	40-60	10-30	100-300	35-38	30-36	35-40
Dense	60-80	30-50	300-500	38-41	36-41	40-45
Very Dense	> 80	> 50	500-800	41-44	> 41	> 45

Estimating Unconfined Compressive Strength from CPT Data

The undrained strength, based on the Q-test (unconsolidated undrained), can be estimated by using the CPT cone resistance [10] from:

$$S_u = \frac{q_c - \sigma_{total}}{N_k} \tag{7.11}$$

where S_u = Undrained cohesive strength
q_c = Measured CPT cone resistance
σ_{total} = In situ total overburden stress
N_k = Empirical undrained strength-bearing factor.

This equation is applicable for most sedimentary, non-sensitive clays. Typical values of N_k are shown in Table 7.7.

Prior knowledge of the *in situ* state of the clay is required when using Table 7.7. The actual value of N_k can be determined by iteration. It is good practice to estimate N_k by first assuming a medium stiff normally consolidated condition (N_k = 13), then calculating S_u using Equation 7.11 and then calculating the S_u/σ'_v ratio, where σ'_v is the effective overburden pressure. If the calculated S_u/σ'_v is 0.29 to 0.33 and S_u is 250 to 500 psf (medium stiff consistency), then the assumed soil condition is correct. If the calculated S_u/σ'_v is too high, then the clay is probably overconsolidated or it has a stiffer consistency, which requires a higher N_k. Iterations are continued until the conditions of Table 7.7 are matched.

Estimating Drained Friction Angle from CPT Data

There are two methods for estimating the drained friction angle of clean sands: (a) correlating ϕ' and normalized cone tip resistance, and (b) predicting the SPT and using it in well-developed correlations between ϕ' and SPT.

Estimation Using Cone Tip Resistance. Terzaghi, Peck, and Mesri [4, Figure 19.5] presented an empirical correlation that indicates

$$\phi' = 28° + 12.4 \log q_{c1e} \tag{7.12}$$

where the normalized tip resistance, q_{c1e}, measured in MPa, is calculated using Equation 7.6. The correlation of Equation 7.12 is applicable mainly to normally consolidated young sand deposits. It underestimates the friction angle of compressible carbonate sands by several degrees and overestimates the friction angle of aged or overconsolidated sands by several degrees.

TABLE 7.7 Typical N_k values for estimating the unconsolidated undrained strength of clays using the CPT cone resistance.

Clay condition	N_k range	N_k average (typical)
Normally consolidated, normal sensitivity, soft to medium stiff	10 to 16	13
Normally consolidated, moderately sensitive, soft to very soft	9 to 13	11
Moderately overconsolidated, non-fissured	15 to 20	17
Highly overconsolidated, fissured	17 to 23	19

Estimation Using SPT Correlations. On those projects where both a number of CPT soundings and several SPT sample borings have been made, it is possible to determine the grain-size distributions for several representative samples of sand. Then the correlation between the mean grain size, D_{50}, and the ratio q_{c1e}/N (measured in tsf) can be determined from Figure 7.22.

A second method involves using Figure 7.23, an overlay on the CPT soil characterization chart of figure 7.19. This method uses both the normalized tip resistance and the normalized friction ratio, but does not require grain size data.

Estimating Pre-consolidation Pressure

A reasonably consistent relationship has been found between the unconsolidated undrained shear strength, or cohesion, and the effective overburden pressure for normally consolidated cohesive soils. This is illustrated in Figure 7.24, which shows (a) the increase in effective vertical pressure with depth in a uniform clayey deposit, (b) the increase in shear strength (cohesion) with depth, and (c) a consistent relationship between the effective vertical pressure and the strength, dependent on the plasticity index of the soil, of a normally consolidated clay. The relationship is shown in Equation 7.13 [18].

$$p' = \frac{c_u}{0.11 + 0.0037\,PI} \tag{7.13}$$

where c_u = Undrained cohesion (one-half of unconfined compressive strength)
p' = Effective overburden pressure
PI = Plasticity index.

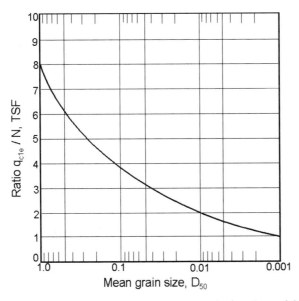

FIGURE 7.22 Relationship between mean grain size, D_{50}, and the ratio q_{c1e}/N (tsf).

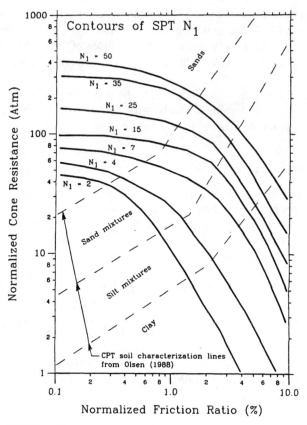

FIGURE 7.23 Estimation of SPT N_1 from normalized CPT tip resistance and friction ratio [10].

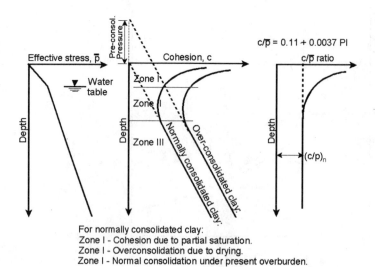

FIGURE 7.24 Relationship between undrained cohesion and effective overburden pressure in a clay layer.

When the soil profile indicates that the effective normal stress is higher than calculated, this is a reasonable indicator that the soil is pre-consolidated, as shown by the right-shifted curve in the center of Figure 7.24. The probable preconsolidation pressure may also be determined by the least-squares line fitted to the data in the deeper section of the profile backward to the origin. The height above the present surface represents the weight that has been removed by erosion.

This relationship can be observed in the data from any of the SPT, VST, or CPT field tests, but it is not observed with tube samples and unconfined compression tests. The reason for this is not fully understood, but it is probably related to the release of the *in situ* lateral confining pressure when the sample is removed from the sampling tube.

Estimation of Liquefaction Potential

As shown in Figure 5.7, dense sands *dilate* during shear and loose sands *contract*. When a loose sand is subjected to a sudden shock force, the sand grains attempt to move into a more stable position. The soil is momentarily sheared in an undrained manner and full pore water pressure is developed, resulting in a complete loss of shear strength. This condition is referred to as *liquefaction* and it occurs most readily in soils of Holocenegeologic age. Liquefaction of pre-Pleistocene deposits is rare.

Liquefaction occurs only in saturated, clean sands and silts having little or no clay content

Fine silts, with platelike particles, and clays have sufficient cohesion to inhibit liquefaction. Saturated gravels existing between finer impermeable layers have liquified.

Cyclic Stress Ratio Method. A common method of evaluating the liquefaction potential of a soil deposit uses the cyclic stress ratio (CSR). In this method, the cyclic shear stress is calculated at all points in the soil profile due to a seismic acceleration applied to the surface. The cyclic shear stress is taken as 0.65 τ (static). The intensity of the cyclic shear stress is reduced by applying a reduction factor to account for depth. The cyclic shear stress is calculated as:

$$\tau_{cyc} = 0.65 \frac{a_{max}}{g} \sigma_v r_d \tag{7.14}$$

where τ_{cyc} = Uniform cyclic shear stress
a_{max} = Peak ground surface acceleration
g = Acceleration of gravity
σ_v = Total vertical stress
r_d = Stress reduction factor (see Figure 7.25).

The procedure for calculating the cyclic shear stress for a soil profile is illustrated in Problem 7.11. Then, either the SPT method or the CPT method can be used to establish whether any part of the soil profile has the potential to liquefy. The Cyclic Stress Ratio is defined as

$$CSR = \frac{\tau_{cyc}}{\sigma'_{v0}} \tag{7.15}$$

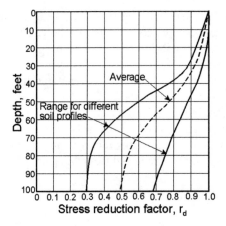

FIGURE 7.25 Reduction factor for estimating cyclic shear stress as a function of depth below level or gently sloping ground surfaces (redrawn from Seed and Idriss [19]).

where CSR = Cyclic Stress Ratio
  τ_{cyc} = Cyclic shear stress
  σ'_{v0} = Initial vertical effective stress.

Using the SPT for Liquefaction Prediction. In most countries, the SPT is commonly used for characterization of liquefaction resistance. Seed, Idriss, and Arango [20] developed a chart (Fig. 7.26) from empirical evidence showing the minimum CSR at which liquefaction could be expected to occur in earthquakes of magnitude 7.5 for sands with various fines contents. The minimum CSR for earthquakes of other magnitudes can be obtained by multiplying the CSR determined by Equation 7.15 by the factors shown in Table 7.8

Using the CPT for Liquefaction Prediction. The normalized cone-tip resistance has been used in conjunction with grain-size tests of soil samples to determine a soil's potential for liquefaction. The CPT is considered to have an advantage over the SPT in its ability to detect thin seams of soil. Olsen and Koester [21] developed an empirical relationship for determining the minimum CSR for liquefaction based on the CPT normalized cone resistance and friction ratio, shown in Figure 7.27.

TABLE 7.8 Magnitude correction factors to be used for with CSR for earthquakes of magnitudes other than 7.5.

Magnitude, M	$CSR_M/CSR_{M=7.5}$
5.25	1.50
6	1.32
6.75	1.13
7.5	1.00
8.5	0.89

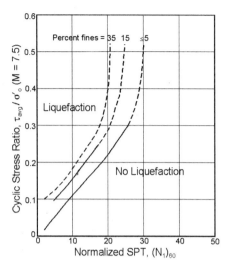

FIGURE 7.26 Minimum CSR for liquefaction compared to the normalized SPT for sands with various fines content in M = 7.5 earthquakes (redrawn from [20])

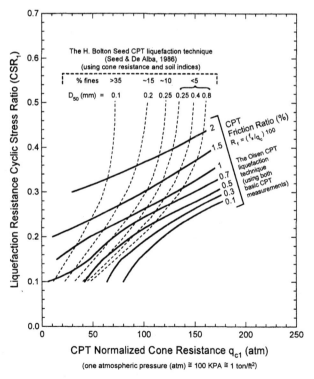

FIGURE 7.27 Liquefaction resistance CSR based on CPT normalized cone resistance and friction ratio (Olsen and Koester [21]) with comparison to the Seed and DeAlba technique based on normalized cone resistance and indices from tests of soil samples.

PLANNING FOR A SUBSURFACE INVESTIGATION

Factors Affecting the Plan for an Investigation

The strategy for a geotechnical engineering subsurface investigation must consider three general factors that establish the type and magnitude of the investigation. The factors are

1. The non-uniformity (variability) in the location, thickness, and extent of soil and rock deposits, and in their engineering properties.
2. The size, loading, and complexity of the structure to be supported by the subsoils and rock. The more complex the structure, the more sensitive it is to distortion; the heavier the structure, the larger and more detailed the investigation must be.
3. The cost and value of additional information. There is no mandatory magnitude or type of subsurface investigation, and the value of additional information decreases with the amount of information available. There is a break-even point where the cost of gathering additional information is more than the savings to the project.

How Many Borings, Where, and How Deep?

The greater the variety of soil types and the variability of the soil properties, the greater the number of borings, samples, and tests needed to achieve a satisfactory level of confidence in the cost effectiveness of the investigation. Conventionally, a uniform spacing of borings with a few samples from each boring is specified. Although this might not be the most economical method of sampling, it has much merit in disclosing changes in soil type.

Boring Layout. General guidance for preliminary and final boring layout is presented in Tables 7.9 and 7.10 according to the type of structure or problem being investigated. Boring layout should also be governed by the geology of the site.

TABLE 7.9 Suggested Spacing of Borings (Adapted from Teng [22] and Sowers [23])

	Distance between borings, feet Horizontal stratification of soil or rock			Minimum number of borings for each structure
	Uniform	Average	Erratic	
Project				
Multi-story buildings	150	100	50	4
One or two-story buildings	200	100	50	3
Bridge piers, abutments, television towers, etc.		100	25	1 - 2 for each foundation unit
Highways	2000-1000	1000-500	200-100	
Borrow pits (for fill)	1000-500	500-200	100-50	
Earth dams, dikes	400-200	200-50	100-25	

Geological Sections. Arrange borings so that geological sections can be determined at the most useful orientations for final siting and design. Borings in slide areas should establish the full geological section necessary for stability analyses.

Critical Strata. Where detailed settlement, stability, or seepage analyses are required, include a minimum of two borings to obtain undisturbed samples of critical strata. Provide sufficient preliminary sample borings to determine the most representative location for undisturbed sample borings.

TABLE 7.10 Guidelines For Boring Layout (from NAVFAC DM-7.01 [24])

Area for Investigation	Boring Layout
New site of wide extent.	Space preliminary borings 200 to 500 feet apart so that the area between any four borings includes approximately 10% of total area. In detailed exploration, add borings to establish geological sections at the most useful orientations.
Development of site on soft compressible strata	Space borings 100 to 200 feet at possible building locations. Add intermediate borings when building site are determined.
Large structure with separate closely spaced footings.	Space borings approximately 50 feet in both directions, including borings at possible exterior foundation walls at machinery or elevator pits, and to establish geological sections at the most useful orientations.
Low-load warehouse building or large area.	Minimum of four borings at the corners plus intermediate borings at interior foundations sufficient to define subsoil profile.
Isolated rigid foundation, 2,500 to 10,000 square feet in area.	Minimum of three borings around perimeter. Add interior borings depending on initial results.
Isolated rigid foundation, less than 2,500 square. feet in area.	Minimum of two borings at opposite corners. Add more for erratic conditions.
Major waterfront structures, such as drydocks.	If definite site is established, space borings generally not more than 50 ft. adding intermediate borings at critical locations, such as deep pumpwell, gate seat, tunnel, or culverts.
Long bulkhead or wharf wall.	Preliminary borings on line of wall at 200 ft. spacing. Add intermediate borings to decrease spacing to 50 ft. Place intermediate borings inboard and outboard of wall line to determine materials in scour zone at toe and in active wedge behind wall.
Slope stability, deep cuts, high embankments.	Provide three to five borings on line in the critical direction to provide geological section for analysis. Number of geological sections depends on extent of stability problem. For active slide, place at least one boring up slope of sliding area.
Dams and water retaining structures.	Space preliminary borings approximately 200 ft. over the foundation area. Decrease spacing on centerline to 100 ft. by intermediate borings. Include borings at the location of cutoff, critical spots in abutment, spillway, and outlet works.

Boring Depths. The depth to which borings should be made depends on the sizes and types of proposed structures (see Tables 7.11 and 7.12). It also is controlled to a great degree by the characteristics and sequence of the subsurface materials encountered.

Unsuitable Foundation Strata. Extend all borings through unsuitable foundation strata, such as unconsolidated fill; peat; highly organic materials; soft, fine-grained soils; and loose, coarse-grained soils in order to reach hard or compact materials of suitable bearing capacity.

Fine-Grained Strata. Extend borings in potentially compressible fine-grained strata of great thickness to a depth where the stress from the superposed load is so small that corresponding consolidation will not significantly influence surface settlement.

Compact Soils. Where stiff or compact soils are encountered at shallow depths, extend borings through this material to a depth where the presence of an underlying weaker stratum cannot affect stability or settlement.

Bedrock Surface. If bedrock surface is encountered and general character and location of rock are known, extend one or two borings 5 feet into sound, unweathered rock. Where the location and character of rock are unknown, or where boulders or irregularly weathered material are likely geologically, increase the number of borings penetrating into rock to bracket the area. In cavernous limestone areas, extend borings through strata suspected of containing solution channels.

TABLE 7.11 Suggested Depth of Borings (Adapted from Teng [22])

Area for Investigation	Boring Depth
Structural foundations.	The depth of borings depends upon the soil profile and the type of feasible foundation. Proceed as follows: (1) If no preliminary soil information is available, start with one or two deep borings to bedrock or to a depth equal to the width of the structure. (2) Analyze the boring results and determine the number and depths of additional borings. Borings should be carried to: (a) below any organic soil, muck, artificial fill, or compressible layer; (b) sufficiently deep for establishing the bottom elevation of foundation (footings, piles, or caissons); and (c) sufficiently deep for checking the possibility of a weaker soil, at a greater depth, which may settle under the sustained load.
Highways and airfields.	Minimum depth of borings is 5 ft but should extend below organic soil, muck, artificial fill, or compressible layers such as soft clays and silts.
Retaining walls and quay walls.	(1) Below organic soil, muck, artificial fill, or any compressible layer; (2) deeper than possible surface of sliding; and (3) deeper than width of the base of wall.
Embankments and cuts.	(1) Below organic soil, muck, artificial fill, or any compressible layer; (2) deeper than possible surface of sliding; and (3) equal to the width at bottom of cuts.

Check Borings. In unfamiliar areas, at least one boring should extend well below the zone necessary for apparent stability, to make sure no unusual conditions exist at greater depth.

Cavernous Limestone. In limestone areas suspected of containing solution channels or cavities, each column location should be investigated. For smaller structures, locate the boring or probe at each planned column location. For large structures and area investigation, use the indirect methods noted below, followed by borings or probes in the final column locations and on close centers (25 ft under walls or heavily loaded areas). Aerial photographs can be used effectively for detecting sinkholes.

TABLE 7.12 Guidelines For Boring Depths (from NAVFAC DM-7.01 [24])

Area for Investigation	Boring Depth
Large structure with separate, closely spaced footings.	Extend to depth where increase in vertical stress for combined foundations is less than 10% of effective overburden stress. Generally all borings should extend to no less than 30 feet below the lowest part of the foundation unless rock is encountered at shallower depth.
Isolated rigid foundation	Extend to depth where increase in vertical stress decreases to 10 percent of bearing pressure. Generally all borings should extend to no less than 30 feet below the lowest part of the foundation unless rock is encountered at shallower depth.
Long bulkhead or wharf wall.	Extend to depth below dredge line between $\frac{3}{4}$ to $1\frac{1}{2}$ times unbalanced height of wall. Where stratification indicates possible deep stability problem, selected borings should reach top of hard stratum.
Slope stability.	Extend to an elevation below the active or potential failure surface and into hard stratum, or to a depth for which failure is unlikely because of the geometry of cross section.
Deep cuts.	Extend to depth between $\frac{3}{4}$ to 1 times the base width of narrow cuts. Where cut is above groundwater in stable materials, depth of 4 to 8 feet below base may suffice. Where base is below groundwater, determine the extent of pervious strata below base.
High embankments.	Extend to depth between $\frac{1}{2}$ and $1\frac{1}{4}$ times horizontal length of side slope in relatively homogeneous foundation. Where soft strata are encountered, borings should reach hard materials.
Dams and water retaining structures.	Extend to depth of $\frac{1}{2}$ base width of earth dams or 1 to $1\frac{1}{2}$ times height of small concrete dams in relatively homogeneous foundations. Borings may terminate after penetration of 10 to 20 feet in hard an impervious stratum if continuity of this stratum is known from reconnaissance.

EXAMPLE PROBLEMS

EXAMPLE 7.1
Which types of sampling and testing methods would be most appropriate for the following types of structure or facility sites?

a. Foundations for building on fine-grained alluvium
b. Foundations for building on glacial till containing lenses of gravel
c. Foundations for building on deep marsh deposit over bedrock
d. Preliminary study of five miles of new roadway
e. Landslide in thin residual soil over weathered rock.

Solutions:

a. For building on alluvium, we need only shallow information. Use SPT borings or CPT soundings; supplement field data with a few undisturbed tube samples for laboratory index properties and consolidation tests.
b. For building on glacial till containing lenses of gravel, use SPT only and design conservatively. CPT soundings will not penetrate gravels. A thin-wall sampling tube might encounter isolated gravel or layer of gravel and might become deformed.
c. Building on deep marsh deposit over bedrock probably will require deep foundations. This investigation will benefit most from CPT sounding to check if a shallow foundation will work. SPT is also appropriate, but might miss thin layers of material. If the character of the marsh deposit is known from previous experience, then only depth-to-rock is needed for piles. In that case, geophysical sounding or even dynamic penetrometer without sampling might suffice.
d. For a preliminary study of five miles of new roadway; the most economical investigation method is some form of geophysical study, either electric resistivity or seismic refraction. This should be accompanied by a number of SPT borings to serve as ground truth.
e. A landslide in thin residual soil over weathered rock probably has moved on or near the surface of the underlying rock. The character of the slide material probably has changed since the slide event. It should be necessary only to locate the probable slide surface, the rock or other hard layer, to permit back-calculation of the probable strength of the slide material at the time of sliding. Use either a dynamic penetrometer with or without sampling (for index properties tests) or even a handheld sounding rod.

EXAMPLE 7.2
A sample of clayey soil is obtained in the SPT split-barrel device. A hand penetrometer test is made that indicates 1.5 tsf. What is the estimated relative consistency of the soil?

Solution:
From Table 7.3, the somewhat disturbed sample would be described as stiff.

EXAMPLE 7.3

An SPT test is made of a silty sand using a safety hammer with a rope and cathead. The test was made at a depth of 15 feet through a 3.5-inch ID hollow stem auger. The SPT device did not have a liner. The measured SPT field N was 12 blows per foot. The dry weight of the soil layer is 110 lbs/ft³ and the water table is below the test depth. What is the normalized $(N_1)_{60}$ for this in situ test?

Solution:
From Table 7.4, the hammer energy correction, $C_e = 0.92$; the rod length correction, $C_l = 0.85$; the sampler correction, $C_s = 1.0$; and the hole diameter correction for a hollow stem auger, $C_d = 1.0$. The effective overburden pressure is (15 ft) (110 pcf) = 1.65 ksf. The overburden correction (Equation 7.5), $C_N = (2/1.65)^{1/2} = 1.10$. The value of the normalized (Equation 7.4) $(N_1)_{60} = 12 \times 1.10 \times 0.92 \times 0.85 \times 1.0 \times 1.0 = 9$ blows/ft.

EXAMPLE 7.4

A CPT sounding indicated, at a given depth, a cone tip resistance of 9600 kPa (96 atm or 96 tsf) and a friction ratio of 3 percent. At that depth, the effective overburden pressure is 130 kPa (1.3 tsf). What is the identification of the soil at that depth?

Solution:
The low friction ratio suggests a sandy soil. The normalized tip resistance is $q_{cle} = 9600/1.3 = 7385$ kPa (73.8 atm). For trial No. 1, using Equation 7.6 with stress exponent $c = 0.60$, $q_{cle} = 73.8 / (1.3)^{0.6} = 63$ the corresponding value for the stress exponent $c = 0.85$. For trial No. 2, using stress exponent $c = 0.75$, $q_{cle} = 73.8 / (1.3)^{0.75} = 61$ the corresponding value for the stress exponent $c = 0.82$. For trial No. 3, using stress exponent $c = 0.80$, $q_{cle} = 73.8 / (1.3)^{0.80} = 60$ the corresponding value for the stress exponent, $c = 0.80$, which matches the assumed value. This indicates a "normally consolidated, medium dense, clean (low fines) sand."

EXAMPLE 7.5

A boring is made in a clay deposit and an SPT is made at a depth of 10 feet. The normalized $(N_1)_{60} = 10$ blows/ft. The SPT sample is tested for Atterberg limits in the laboratory. What is the best estimate of the unconfined compressive strength of the clay if it classifies in the USCS as (a) clay of low plasticity, SC-ML? (b) clay of medium plasticity, CL? (c) clay of high plasticity, CH?

Solutions:
  a. For a clay of low plasticity, Figure 7.21 indicates $q_u = 75$ kPa = 0.75 tsf.
  b. For a clay of medium plasticity, Figure 7.21 indicates $q_u = 150$ kPa = 1.5 tsf.
  c. For a clay of high plasticity, Figure 7.21 indicates $q_u = 250$ kPa = 2.5 tsf.

EXAMPLE 7.6

Estimate the drained friction angle, ϕ' for the silty sand of Problem 7.3.

Solution:
Using Equation 7.10, $\phi' = 28° + 9 / 4 = \sim 30°$.

EXAMPLE 7.7
A CPT sounding was made in a clay layer. The cone-tip resistance at a given depth was 2270 kPa and at the same depth the total overburden pressure was 570 kPa. Estimate the undrained shear strength (one half of unconfined compressive strength) for the clay.

Solution:
Assume initially that the clay is medium stiff, with $N_k = 13$.
Using equation 7.11, $S_u = (2270 - 570)/13 = 130$ kPa.
Table 7.3 indicates a clay soil having a very stiff consistency.
Try again, using $N_k = 17$; $S_u = (2270 - 570)/17 = 100$ kPa.
This is consistent ($q_u = 200$ kPa) with a stiff consistency in Table 7.3.

EXAMPLE 7.8
For the soil tested by the CPT sounding described in Problem 7.4, estimate the drained friction angle.

Solution:
Two methods are available with the information given:

a. Using Equation 7.12, with a normalized cone tip resistance equal to $q_{c1e} = 73.8$ tsf = 7.38 MPa, then $\phi' = 28° + 12.4 \log 7.38 = \sim 39°$. As indicated in the discussion following Equation 7.12 this method "overestimates the friction angle of aged or overconsolidated sands by several degrees."

b. Entering Figure 7.23 with the values of $q_{c1e} = 73.8$ tsf (73.8 atm) and friction ratio = 3%, the indicated $N = 8$ blows/ft. Equation 7.9 (for clean sand) then indicates that $\phi' = 30° + N/3 = \sim 33°$.

EXAMPLE 7.9
Assume that the clay of Problem 7.7 exists at a depth such that the effective overburden pressure is 300 kPa and that an Atterberg limits test of a sample of the clay indicated PI equals 30. What is the overconsolidation ratio (OCR) for the clay?

Solution:
By using Equation 7.13, we find that the effective overburden pressure for the measured undrained shear strength is $p' = 100$ kPa $/ (0.11 + 0.0037 \times 30) = 452$ kPa. The actual effective overburden pressure is 300 kPa. Therefore, OCR $= 452 / 300 = 1.50$. This is consistent with the stiffness of the clay.

EXAMPLE 7.10
A site has the following soil profile: (a) 0 to 12 ft, silty sand with 35 percent fines, $\gamma_{d.} = 108$ pcf; (b) 12 to 20 ft, silty sand with 15 percent fines, $\gamma_{d.} = 112$ pcf; (c) 20 to 40 ft, clean sand with 4 percent fines, $\gamma_{d.} = 116$ pcf; and (d) the water table is 10 ft below the surface. The site is subjected to an earthquake that produces a maximum peak acceleration of 0.30 g. Estimate the maximum shear stress and compute the variation of cyclic shear stress with depth.

Solutions:
At 10 ft:
$$\sigma_v = (10 \text{ ft})(108 \text{ pcf}) = 1080 \text{ pcf.}; r_d = 0.98 \text{ (from Figure 7.25)}.$$
$$\tau_{cyc} = 0.65 (0.3) 1080 (0.98) = 206 \text{ lb/ft}^2 \text{ (from Equation 7.14)}.$$
$$CSR_{M=7.5} = 206 / 1080 = 0.19 \text{ (from Equation 7.15)}.$$

At 20 ft:
$$\sigma_v = 1080 + (2 \text{ ft})(108) + (8 \text{ ft})(112 \text{ pcf}) = 2192 \text{ pcf}; r_d = 0.96.$$
$$\tau_{cyc} = 0.65 (0.3) 2192 (0.96) = 410 \text{ lb/ft}^2$$
$$CSR_{M=7.5} = 410 / [1080 + 2 \times (108-62.4) + 8 \times (112-62.4)] = 0.26.$$

At 30 ft:
$$\sigma_v = 2192 + (10 \text{ ft})(116 \text{ pcf}) = 3368 \text{ pcf}; r_d = 0.94.$$
$$\tau_{cyc} = 0.65 (0.3) 3368 (0.94) = 614 \text{ lb/ft}^2$$
$$CSR_{M=7.5} = 614 / [1568 + 10 \times (116-62.4)] = 0.29.$$

At 40 ft:
$$\sigma_v = 3368 + (10 \text{ ft})(116 \text{ pcf}) = 4528 \text{ pcf}; r_d = 0.86.$$
$$\tau_{cyc} = 0.65 (0.3) 4528 (0.86) = 759 \text{ lb/ft}^2$$
$$CSR_{M=7.5} = 759 / [2104 + 10 \times (116-62.4)] = 0.29.$$

EXAMPLE 7.11

For the soil profile of Problem 7.10, determine the factor of safety against liquefaction if the following normalized SPT N-values were measured: (a) at 10 ft, $N = 13$ blows per foot; (b) at 20 ft, $N = 16$ bpf; (c) at 30 ft, $N = 22$ bpf; and (d) at 40 ft, $N = 30$ bpf.

Solutions:
  a. At 10 ft, for $N = 13$, the minimum $CSR_{M=7.5} = 0.23$.
     Factor of Safety (FS) = 0.23/0.19 = 1.2. No liquefaction is expected.
  b. At 20 ft, for $N = 16$, the minimum $CSR_{M=7.5} = 0.23$.
     FS = 0.23/0.26 = 0.9. Liquefaction is expected.
  c. At 30 ft, for N = 22, the minimum $CSR_{M=7.5} = 0.24$.
     FS = 0.24/0.29 = 0.8. Liquefaction is expected.
  d. St 40 ft, for $N = 30$, the minimum $CSR_{M=7.5} = 0.45$.
     FS = 0.45/0.29 = 1.6. No liquefaction is expected.

EXAMPLE 7.12

For the soil profile of Problems 7.10 and 7.11, (a) determine the factor of safety for the soil at the 10 ft depth if the earthquake had a magnitude of 8.5; (b) determine the factor of safety for the soil at the 20 ft depth if the earthquake had a magnitude of 6.5.

Solutions:

a. From Table 7.8, $CSR_{M=8.5} / CSR_{M=7.5} = 0.89$. The minimum CSR at 10 ft then becomes $0.89 \times 0.23 = 0.20$. FS = 0.20/0.19 = 1.05. Although liquefaction is not expected, it is possible because of statistical scatter in the basic data.

b. From Table 7.8, $CSR_{M=6.5} / CSR_{M=7.5} = 1.2$. The minimum CSR at 20 ft then becomes $1.2 \times 0.23 = 0.28$. FS = 0.28/0.26 = 1.08. No liquefaction is expected.

REFERENCES

1. U. S. Army. 1995. "Geophysical Exploration for Engineering and Environmental Investigations," pp. 2-3, Engineer Manual, EM 1110-1-1802. U.S Army Engineers, Washington, DC.
2. Transportation Research Board. 1996. "Landslides: Investigation and Mitigation." A.K Turner and R.L. Schuster, eds., *Special Report 247*. Transportation Research Board, National Research Council. Washington, DC.
3. Redpath, B. B. 1973. "Seismic Refraction Exploration for Engineering Site Investigations." Technical Report E-73-4. U.S. Army Engineer Waterways Experiment Station. Vicksburg, MS.
4. Terzaghi, K., Peck, R. B., and Mesri, G. 1996. *Soil Mechanics in Engineering Practice*. 3rd edition. New York: John Wiley & Sons.
5. ASTM. 1999. "Natural Building Stones; Soil and Rock; Geotextiles," *1999 Annual Book of ASTM Standards*, Vol. 04.08. American Society for Testing and Materials. Philadelphia, PA.
6. Hvorslev, M. J. 1949. "Subsurface Exploration and Sampling of Soils for Civil Engineering Purposes." U.S. Army Engineer Waterways Experiment Station, Vicksburg, MS (republished in 1965 by the Engineering Foundation, United Engineering Center, New York).
7. U.S. Army. 1986. "Shear Strength of Soil and Rock." Lecture notes, Corps of Engineers Training Course, Geotechnical Laboratory. U.S. Army Engineer Waterways Experiment Station, Vicksburg, MS.
8. Skempton, A.W. 1986. "Standard Penetration Test Procedures and the Effects in Sands of Overburden Pressure, Relative Density, Particle Size, Aging, and Overconsolidation." *Gèotechnique*. Vol. 36 No. 3.
9. Olsen, R.S. and Mitchell, J.K. Oct. 1995. "CPT Stress Normalization and Prediction of Soil Classification." *Proceedings of the International Symposium on Cone Penetration Testing*. Linkoping, Sweden.
10. Olsen, R.S. 1994. "Normalization and Prediction of Geotechnical Properties Using the Cone Penetrometer Test (CPT)," Technical Report GL-94-29. U.S. Army Engineer Waterways Experiment Station, Vicksburg, MS.
11. Personal Communication. (1995). Dr. Richard S. Olsen, research civil engineer. U.S. Army Engineer Waterways Experiment Station, Vicksburg, MS, June.
12. U. S. Army. 1992. "Bearing Capacity of Soils." Engineer Manual, EM 1110-1-1905, Table 3-1. U.S Army Engineers, Washington, DC.
13. Schmertmann, J. H. 1978. "Guidelines for Cone Penetration Test Performance and Design," Report No. FHWA-TS-78-209. U.S. Department of Transportation, Federal Highway Administration. McLean, VA.
14. Terzaghi, K. and Peck, R.B. 1967. *Soil Mechanics in Engineering Practice*. 2nd edition. New York: John Wiley & Sons.

15. Meyerhof, G.G. 1974. "Penetration Testing Outside Europe: General Report." *Proceedings of the European Symposium on Penetration Testing*, Vol. 2.1. National Swedish Institute for Building Research. Gavleä, Sweden:
16. Peck, R.B., Hanson, W.E., and Thornburn, T.H. 1974. *Foundation Engineering*, 2nd edition. New York: John Wiley & Sons.
17. Meyerhof, G.G. 1974. "Ultimate Bearing Capacity of Footings on Sand Overlying Clay." *Canadian Geotechnical Journal*. Vol. 11.
18. Skempton, A. W. 1948. "Vane Tests in the Alluvial Plain of the River Forth near Grangemouth." *Gèotechnique,* Vol 1, No. 2.
19. Seed, H. B. and Idriss, I. M. 1971. "Simplified Procedure for Evaluating Soil Liquefaction Potential." *Journal of the Soil Mechanics and Foundations Division*. Vol. 107, No. SM9.
20. Seed, H. B., Idriss, I.M., and Arango, I. 1983. "Evaluation of Liquefaction Potential Using Field Performance Data." *Journal of Geotechnical Engineering*. Vol. 109, p. 458.
21. Olsen, R.S. and Koester, J.P. Oct. 1995. "Prediction of Lquefaction Resistance Using the CPT." *Proceedings of International Symposium on Cone Penetration Testing*. Linkoping, Sweden.
22. Teng, W.C. 1962. *Foundation Design*. Englewood Cliffs, NJ: Prentice-Hall.
23. Sowers, G.F. 1979. *Introductory soil Mechanics and Foundations: Geotechnical Engineering*; Fourth Edition. New York: Macmillan Publishing.
24. NAVFAC DM 7.01. 1982. "Soil Mechanics", Design Manual 7.01," Naval Facilities Engineering Command, Department of the Navy. Alexandria, VA. September.

CHAPTER 8
SHALLOW FOUNDATIONS—FOOTINGS AND RAFTS

The concentrated load from a building column or wall ultimately must be supported by the underlying soil or rock without excessive settlement. Because the column or wall load divided by the contact area is invariably much greater than the supporting capacity of the soil, a punching shear failure is likely to occur, permitting severe distortion of the structure. To prevent this, it is common to use either a foundation element having a bearing area sufficiently large to reduce the shear stresses in the soil to an acceptable level, or a deep foundation to transfer the load to a more competent layer at depth, or some combination of the two.

The design of a foundation must consider two somewhat independent factors: *bearing capacity* and *settlement*. This chapter deals with shallow foundation elements, those in which the depth of the foundation below the ground surface is less than the width of the foundation. Chapter 9 deals with deep foundations.

BEARING CAPACITY

Bearing capacity is the ability of the supporting soil or rock to sustain the load from any engineered structure without a shear failure. Bearing pressures that exceed the shearing resistance of the soil result in large downward movements of the structure, often in a tilting mode. Shear failure movements are typically 0.5 feet to more than 10 feet (15 cm to 3 m) in magnitude.

SETTLEMENT

Applying a bearing pressure that is safe with respect to shear failure does not ensure that settlement of the foundation will be within acceptable limits for the structure. Therefore, it is also necessary to evaluate the settlement to be expected and to proportion the foundation elements accordingly. When sufficient data are not available for a valid settlement analysis it often is sufficient to apply a suitable factor of safety to the ultimate bearing capacity so that the resulting pressure will be below the preconsolidation pressure of the soil and only an acceptably small settlement will occur.

SHEAR FAILURE MODES

Foundation-bearing capacity failures can be grouped into three categories, depending on the compressibility of the supporting soil: general shear, punching shear, and local shear.

General Shear Failure

As shown in Figure 8.1, a general shear failure involves a total rupture of the underlying soil. The shear surfaces occupy a triangular zone immediately below the footing; this zone moves downward as part of the foundation due to friction with the concrete. The soil outside this triangle moves laterally and upwardly, causing a bulge at the ground surface. This type of failure requires soil that is incompressible, such as dense sand or hard clay. Failure is catastrophic. The load-settlement curve for a footing shows a distinct rupture load at which all parts of the shear surfaces reach the ultimate shear strength of the soil.

Punching Shear Failure

As shown in Figure 8.2, a punching shear failure does not develop the shear planes of a general shear failure. Soils that are highly compressible, such as loose sand and loess, will compress directly under the footing and will develop a vertical shear plane between soil and concrete at the perimeter of the footing. There is very little lateral soil displacement and the unit compression of the soil within the footing perimeter decreases with depth. Larger footings have greater settlement than smaller footings at the same unit pressure. Failure is indistinct and not catastrophic.

Local Shear Failure

As shown in Figure 8.3, local shear failure is characterized by a slip path that is not well defined except immediately beneath the footing. It most likely will occur in loose sands, silty sands, and weak clays. An increase in the applied load causes increasing deformation, but failure is not catastrophic.

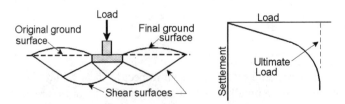

FIGURE 8.1 General shear failure of footing. Load settlement curve displays a vertical line at the ultimate shear strength of the soils on the shear surfaces.

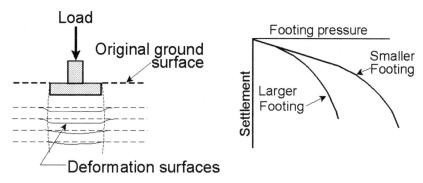

FIGURE 8.2 Punching shear failure of footing. Load settlement curve does not reach a vertical line. [1]

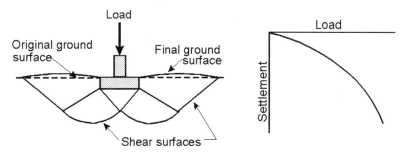

FIGURE 8.3 Local shear failure of footing. Load settlement curve does not reach a vertical line.

THEORETICAL BEARING CAPACITY EQUATIONS

In 1943, Terzaghi [2] published the following bearing-capacity equation for shallow continuous footings for the general shear-failure mode, corresponding to a dense or hard soil mass located above the water table:

$$q_d = cN_c + \gamma D_f N_q + 0.5\gamma B N_\gamma \tag{8.1}$$

where q_d = Ultimate bearing capacity
  c = Cohesion, based on the Coulomb shear-strength equation
  γ = Unit weight of soil
  D_f = Depth of foundation below the outside ground surface
  B = Width of foundation element
N_c, N_q, N_γ = Bearing capacity factors (see Figure 8.4).

The bearing capacity factors in Equation 8.1 also were developed by Terzaghi [2]. The values of N in Figure 8.4 were developed by Meyerhof [3] using advanced procedures.

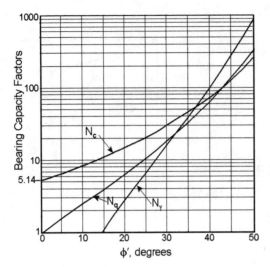

FIGURE 8.4 Relationship between bearing capacity factors and effective friction angle, ϕ, Terzaghi [2] with N_γ factors after Meyerhof [3].

Footings of Finite Length

Based on numerical analyses and on experiments, two semi-empirical equations have been developed [4] for footings of finite length. For a circular footing with radius R on dense or hard soil,

$$q_{dr} = 1.2cN_c + \gamma D_f N_q + 0.6\gamma R N_\gamma \qquad (8.2)$$

and for a square footing, $B \times B$, on dense or hard soil,

$$q_{dr} = 1.2cN_c + \gamma D_f N_q + 0.4\gamma B N_\gamma \qquad (8.3)$$

Footings with Inclined Load

Meyerhof [5] published solutions to the problems of an inclined load on horizontal footings and an axial load on an inclined footing. Graphical solutions are included in NAVFAC DM-7.02 [6] and are reproduced in Figure 8.5. The N_q factor has been integrated into the N_c and N_γ factors as shown in the equations at the bottom of Figure 8.5 (see Equations 8.4 and 8.5 below).

Shallow Footings on or Near Slope

Meyerhof [7] also published solutions to the problems of continuous footings on or near a slope. Key figures for solving the equations are shown in Figure 8.6. Charts for determining the N_{cq} and $N_{\gamma q}$ factors, which also are included in NAVFAC DM-7.02 [6], are reproduced in Figure 8.7.

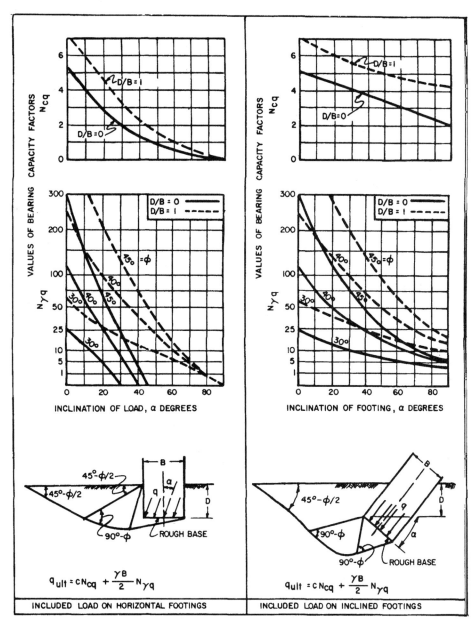

FIGURE 8.5 Ultimate bearing capacity of continuous footings with inclined loads (NAVFAC DM-7.02 [6] after Meyerhof [5]).

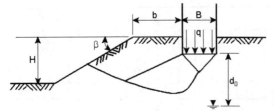

CASE I - CONTINUOUS FOOTING AT TOP OF SLOPE

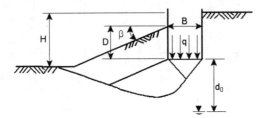

CASE II - CONTINUOUS FOOTING ON SLOPE

FIGURE 8.6 Keys to the use of Figure 8.7 for shallow continuous footings placed on or near a slope (adapted from NAVFAC DM-7.02 [6]).

Case I—Continuous Footing at Top of Slope. This technique, based on the procedure in NAVFAC DM-7.02 [6], involves using the left side of Figure 8.7.

1. Identify the location of the water table with respect to the base of the footing, d_0. Then,

2. If water at $d_0 \geq B$
$$q_{ult} = cN_{cq} + 0.5B\gamma N_{\gamma q} \qquad (8.4)$$

   Water at ground surface
$$q_{ult} = cN_{cq} + 0.5B\gamma_{sub} N_{\gamma q} \qquad (8.5)$$

   Use 0.4B for square footings or 0.6R for round footings instead of 0.5B in Equations 8.4 and 8.5.

3. If $B \leq H$, obtain N_{cq} from the lower part of Figure 8.7 for stability number $N_0 = 0$ and interpolate for values of $0 < D/B < 1$ to find $N_{\gamma q}$ in the upper part of Figure 8.7.

4. Interpolate q_{ult} between Equations 8.4 and 8.5 for water at a level between ground surface and $d_0 = B$.

5. If $B > H$, obtain N_{cq} from the lower part of Figure 8.7 with stability number $N_0 = \gamma H/c$. Interpolate for values $0 < D/B < 1$ for $0 < N_0 < 1$. If $N_0 \geq 1$, the stability of the slope controls the ultimate bearing pressure.

6. Interpolate q_{ult} between Equations 8.4 and 8.5 for water at a level between ground surface and $d_0 = B$. For water at ground surface and sudden drawdown, substitute ϕ' for ϕ in Equation 8.5.

7. For cohesive soil ($\phi=0$), substitute D for B/2 and use $N_{\gamma q} = 1$ in Equation 8.4.

Case II—Continuous Footing on a Slope. Use the same criteria used for Case I, except that N_{cq} and $N_{\gamma q}$ are obtained from the diagrams in Figure 8.7.

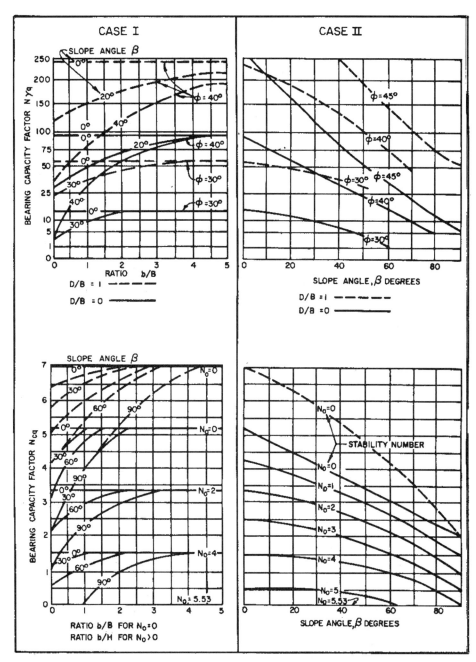

FIGURE 8.7 Bearing capacity factors for shallow continuous footings placed on or near a slope (NAVFAC DM-7.02 [6]).

Bearing Capacity of Cohesive Soils

Single Cohesive Layer. The ultimate bearing capacity of cohesive soils, mainly clay and plastic silt, which are governed by condition that $\phi = 0$ at the end of construction, is dependent primarily on the Q-test shear strength. For the case of clays meeting the criterion that $\phi = 0$, the bearing capacity factors become $N_\gamma = 0$ and $N_q = 1$, as shown in Figure 8.4. Equation 8.1 is reduced to

$$q_d = cN_c + \gamma D_f \tag{8.6}$$

For those cases in which the increase in bearing capacity per unit of area produced by the surcharge γD_f is exactly compensated by the weight of soil removed for construction of the footing, it is more convenient to deal with the *net bearing capacity*, given by the following equation:

$$q_{d(net)} = q_d - \gamma D_f = cN_c \tag{8.7}$$

Skempton [8] developed relationships for the value of the shape factor, N_c, to be used in Equations 8.6 and 8.7. Although originally presented in graph form, they also can be represented in equation form.

1. For a continuous footing, for $D_f/B \leq 4$

$$N_c = 5.14 + \frac{D_f/B}{0.37 + 0.35\,(D_f/B)} \tag{8.8}$$

2. For a circular or square footing, for $D_f/B \leq 4$

$$N_c = 6.2 + \frac{D_f/B}{0.45 + 0.24\,(D_f/B)} \tag{8.9}$$

3. For a rectangular footing,

$$N_c = \left(0.84 + 0.16\frac{B}{L}\right) N_c\,(\text{square}) \tag{8.10}$$

Two-Layer Cohesive Soil. The shape-related bearing capacity factor for clays N_c was developed by Button [9] for the case in which the cohesive soil ($\phi = 0$) exists as two layers. Button's figures were reproduced in NAVFAC DM-7.2 [6] and are included herein as Figures 8.8 and 8.9. Figure 8.8 deals with the case in which the cohesion values for the two layers are constant within each layer. Figure 8.9 treats the case in which the cohesion values for the two layers vary uniformly within each layer. The value N_c is used in Equations 8.8, 8.9, and 8.10.

Settlement Limitations

Theoretically, no damage will ensue to a structure if it settles uniformly. If, as is usually the case, differential settlement occurs, however, the structure can be damaged. The expected settlement of individual foundation elements should be checked and the allowable bearing pressure based on bearing capacity adjusted accordingly. Table 8.1, based on both theory and observations of structures [10], can be used as a guide.

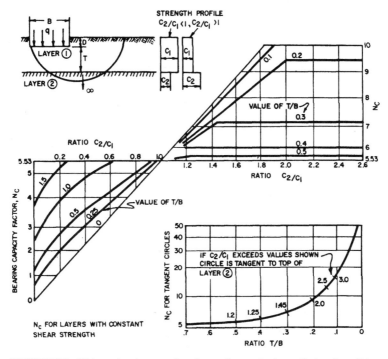

FIGURE 8.8 Ultimate bearing capacity of a two-layer cohesive soil, $\phi = 0$ condition, strength constant with depth (NAVFAC DM-7.02 [6] after Button [9]).

TABLE 8.1 Maximum Permissible Settlement [10].

Type of Movement	Limiting Factor	Maximum Settlement
Total settlement	Drainage and access	0.15 to 0.6 m (0.5 to 2 ft)
	Probability of differential settlement	
	Masonry walls	25 to 50 mm (1 to 2 in.)
	Framed buildings	50 to 100 mm (2 to 4 in.)
Tilting	Tower, stacks	0.004 Base width
	Rolling of trucks, stacking of goods	0.01 Column spacing
	Crane rails	0.003 Column spacing
Curvature	Brick walls in buildings	0.0005-0.002 Column spacing
	Reinforced concrete building frame	0.003 Column spacing
	Steel building frame, continuous	0.002 Column spacing
	Steel building frame, simple	0.005 Column spacing

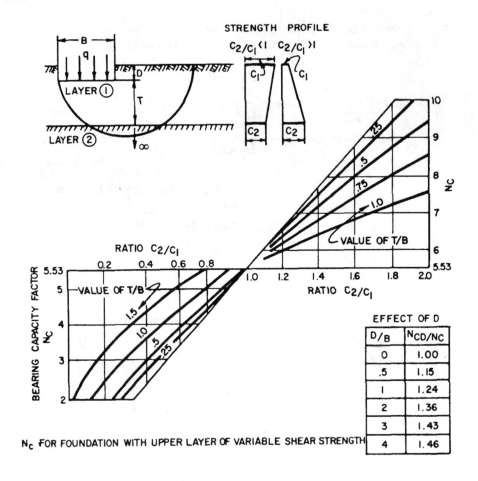

FIGURE 8.9 Ultimate bearing capacity of two-layer cohesive soil, $\phi = 0$ condition, strength varying with depth (NAVFAC DM-7.02 [6] after Button [9]).

BEARING CAPACITY FROM IN SITU TESTS

Full-scale, *in situ* load tests of full-size foundations are rarely done, except for piles and drilled shafts, because of the large loads and the great expense involved. The most common methods for estimating bearing capacity of a soil include the *in situ* tests described in Chapter 7, such as plate bearing, SPT, CPT, and VST, and the laboratory testing of undisturbed samples taken with thin-wall tube samplers.

An advantage of making less precise estimates of bearing capacity by obtaining and testing a large number of samples, instead of making costly full-size load tests, is that greater coverage of the project site is possible. The larger number of field tests also can disclose areas of looser or softer soils that were not expected.

Footings on Sand and Nonplastic Silt

Foundations placed on sand, except for very loose sands or metastable sands, are invariably safe against bearing capacity failure when using normal foundation pressures. Settlement usually will be the governing factor, and for conventional foundation pressures, the footing rarely will settle enough to cause distress in the supported structure. Sand deposits, even those under the most constant geologic conditions, vary significantly and erratically over even a fairly small building site. Therefore, in light of the heterogeneity of sand deposits and the feasibility of providing an adequate factor of safety (typically two) against bearing capacity failure, approximate procedures for proportioning footings for settlement are the best that can be used in practice.

Plate Load Test.

Early applications of soil mechanics to settlement prediction for footings on sand indicated a parabolic relationship between the settlement of a 1 ft (30 cm) square plate and a larger footing. For a large footing the ratio of unit pressures, at equal settlements, approached four. It is now recognized [4] that this ratio, however, was based on an oversimplification of a complex phenomenon and that the results of plate load testing should not be relied upon for foundations on sand. Furthermore, plate load tests are costly and a greater amount of useful information can be obtained by *in situ* testing, such as the SPT and CPT, and by the application of empirical relationships.

SPT Method.

Terzaghi, Peck, and Mesri [4] recommend a semi-empirical approach based on a broad database of field studies of structure settlement on sand. Most of the studies were made using the SPT (Chapter 7) to estimate the relative density of the sand.

The settlement for a given pressure increases with footing width. To ensure that the largest footing will not settle more than 25 mm (1 in.), the *average* settlement for all footings should be limited by about 2/3, or to 16 mm (0.65 mm). Then, using an average corrected SPT value for a normally consolidated sand within a significant depth, B, below the footing,

$$q = S_c \frac{1.4 \bar{N}_{60}}{1.7 B^{0.75}} = 16 \frac{1.4 \bar{N}_{60}}{1.7 B^{0.75}} = 16Q \qquad (8.11)$$

where q = Allowable bearing pressure, in kPa, to limit average settlement
S_c = Settlement, in mm, at end of construction and under permanent live load
$\overline{N}_{60}$ = Average corrected SPT value, in blows per 30 cm (1 ft), in depth $B^{0.75}$
B = Width of footing, in m
Q = Allowable bearing pressure, in kPa, for 1 mm of settlement.

If the sand has been subjected to an effective stress preconsolidation pressure, σ', either geologically or by recent excavation, then Equation 8.11 becomes

$$q > \sigma' \quad q = 16Q + 0.67\sigma' \tag{8.12}$$

$$q < \sigma' \quad q = 3 + 16Q \tag{8.13}$$

For rectangular footings, with length L greater than B, Equation 8.11 becomes

$$S_c(L/B > 1) = S_c(L/B = 1) \left[\frac{1.25\,(L/B)}{(L/B) + 0.25} \right]^2 \tag{8.14}$$

CPT Method. The CPT has not been used extensively for the development of the database used in the empirical SPT method. Therefore, rather than resort to quasi-empirical methods involving estimation of strain, it is considered more practical to simply use the methods of Chapter 7 to estimate SPT values from the CPT and then use the method given above involving Equations 8.11 through 8.14.

Rafts on Sand and Nonplastic Silt

When the area of the footings needed to support a structure exceeds about one-half of the total structure area, it is often desirable to combine all footing loads onto a single mat or raft foundation. The factor of safety of such a large foundation on a deep sand deposit is so great that it does not even require calculation. The design of a raft foundation is governed by settlement.

Experience has shown that if a uniformly loaded area is founded at least 3 m (10 ft) below the surface $(D_f \geq 3\text{ m})$, and the sand layer has a thickness greater than $B^{0.75}$, the settlement will be rather even. Most rafts can easily sustain a total settlement of up to 50 mm (2 in.). Assuming the loads on the raft are well distributed, the differential settlement of a raft is about one-half that of a footing-supported structure. Therefore, the allowable soil pressure for a raft on sand can be assigned a value of twice that given by Equation 8.11.

If the depth of the raft is smaller than 3 m (10 ft), then outward movement of sand at the perimeter may allow settlement of the edges of the raft unless the sand is prevented from yielding within a depth of 2 to 3 m (6 to 9 ft) from the surface. If the depth to bedrock or other hard stratum is less than $B^{0.75}/2$ the allowable soil pressure can be increased [4]. If the sand layer is of limited thickness and is underlain by a clay layer, however, because the raft pressure does not greatly diminish with depth, compression should be evaluated.

Footings on Clay and Plastic Silt

In practice, the cohesion, c, is normally determined by either plate load tests, SPT, CPT, VST, or unconfined compression tests of thin-wall tube samples. When a suitable factor of safety against shear failure (normally 3 for clays) is used, the allowable bearing pressure is generally less than the preconsolidation pressure of the clay by a factor of at least 1.5 to more than 3. This is based on Equation 4.2 and is valid for soils with plasticity index values between 5 and 60 and possibly higher. Therefore, if the footings are designed for a safety factor of 3 and the footings are far enough apart so that the action of each is independent of the others, the differential settlements due to compression of the clay immediately beneath the footings is not likely to exceed 0.75 in. (2 cm), unless the soil is a soft, normally loaded clay.

Plate Load Test. The plate load test can indicate the undrained shear strength for the soil within its significant depth, provided the following conditions are met:

1. If the shear strength is constant within the significant depth, and the width of the actual foundation is less than 4 × the width of the plate, the shear strength determined from the plate test equals the shearing resistance of the soil under the foundation.
2. If the shear strength increases linearly with depth, and the width of the actual foundation is less than 4 × the width of the plate, the shear strength determined from the plate test equals the shearing resistance of the soil under the foundation times the ration of the widths of the foundation and the plate.
3. Plate load testing, however, is expensive and time-consuming compared to other methods, and it is not an indicator of expected settlement. It can give an invalid result (see Chapter 7).

Estimates of Unconfined Compressive Strength. Reasonable estimates of the unconfined compressive strength of clay can be made by any of the methods discussed in Chapter 7. The most commonly used are the SPT, CPT, the VST, and the unconfined compression test of a thin-wall tube sample. Each of the test methods has advantages and physical limitations, as discussed in Chapter 7. The resulting unconfined compressive strength then can be used with any of the theoretical bearing capacity equations given above, especially Equations 8.6 through 8.10.

The selection of the best method for characterizing the *in situ* undrained shear strength of the clay deposit depends on several factors, including the balancing of the higher precision and cost of tube sample tests against the lower precision and cost of field tests. Each of the methods uses a sample to characterize the universe (the clay deposit) from which it was taken. It is essentially a matter of comparing the cost of obtaining the required number of samples or tests for each method to yield the same statistical significance level. A large number of less precise tests is often as good or better than a small number of more precise tests.

Rafts on Clay and Plastic Silt

The factor of safety against a bearing capacity failure on clay, as shown by Equation 8.6, is independent of footing size. There have been a few dramatic failures of large structures founded on clay, mainly grain storage tanks or bins. This led Terzaghi, Peck, and Mesri

[4] to state, "To avoid the risk of such a failure, a raft foundation on clay should be designed so that the excess load divided by the loaded area does not exceed about one-third the value of $q_{d(net)}$ determined by [Equation 8.7]."

If the raft foundation is located beneath a basement and supports a structure, it constitutes with the basement a large hollow footing. The bearing capacity increases with depth by an amount D_f so that, in some cases, a major portion of the total load of the structure is compensated by the weight of the soil removed and the net bearing pressure is greatly reduced. This concept of a *compensated foundation* has been used to produce equal foundation pressures and equal settlement under parts of buildings having unequal heights or unequal total weights, by having the basement for each section at a different depth.

Earthquake Loading of Foundations

Earthquake loading causes additional axial load on the foundation, above the static load. Because of the short time interval in which the maximum acceleration of the structure exerts the additional load, it is common to increase the allowable bearing pressure under the earthquake-induced load by a factor of one third.

Modulus of Subgrade Reaction

Soil is not an elastic medium. The concept of a modulus of subgrade reaction stems from the attempt to make the early part of the stress-strain, or load-settlement, curve for inherently inelastic soil into an elastic modulus. The modulus of subgrade reaction is used in several applications, including the design of rigid pavements, the calculation of immediate settlement under foundations, the design of continuous footings and rafts (beam on elastic foundation), and the evaluation of the deformation of piles under lateral loading.

As shown in Figure 8.10, the modulus is defined as the slope of the load-deflection curve and is similar in all respects to the definition of modulus of elasticity for an inelastic material. Bowles [11] discussed extensively the relation of the modulus of subgrade reaction and the elastic modulus of soils and tabulated typical values, as shown in Table 8.2. The subgrade modulus is defined as

$$k_s = \frac{q}{\delta} \tag{8.15}$$

where k_s = Modulus of subgrade reaction, in force per unit of volume
q = Load or pressure, in force per unit of area
δ = Deflection or strain, in unit of length.

Plate Load Test. The modulus of subgrade reaction can be determined by plate load test. NAVFAC DM-7.01 [12] shows a procedure for interpreting the modulus of subgrade reaction for a 1 ft (30 cm) square plate. The data from the load-settlement curve is plotted on a log-log graph and the yield point is determined where there is a break in the slope of the curve. The subgrade modulus is defined by using the load at one half the yield point and the corresponding settlement. This is essentially the definition of a *tangent modulus*, as shown for the term k_{s-1} in Figure 8.10.

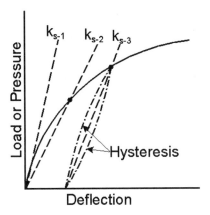

FIGURE 8.10 Definitions of modulus of subgrade reaction, k_s.

Pavement Design. A plate load test procedure for use in the design of airport and highway pavements is specified in ASTM Standards D 1195 and D 1196 [13]. A stack of three circular steel plates, ranging in diameter from 12 in. (30 cm) on top to 30 in. (76 cm) at the bottom, is used to provide rigidity. The stack of plates is placed directly on the subgrade soil. A gradually increased load is applied and the corresponding deflection recorded. The plate pressure at 0.05 in. (1.25 mm) is used to establish the subgrade modulus as shown for the term $k_{s\text{-}2}$ in Figure 8.10 for use in Equation 8.10. This is essentially a *secant modulus*. The modulus is expressed in units of force per unit of volume, such as pounds per square inch per inch, which yields pounds per cubic inch.

Cyclic Loading. Weissmann [14] discussed a procedure for determining the elastic deformation of the foundation of a rotating radar antenna for which small foundation movements are of great concern. The field testing used a test arrangement of stacked circular plates in

TABLE 8.2 Range of Values of Modulus of Subgrade Reaction, k_S [11].

Soil	k_s, kip/ft^3	k_s, kN/m^3
Loose sand	30-100	4800-16000
Medium dense sand	60-500	9600-80000
Dense sand	400-800	64000-128000
Clayey medium dense sand	200-500	32000-80000
Silty medium dense sand	150-300	24000-48000
Clayey soil:		
$\quad q_u < 200$ kPa	75-150	12000-24000
$\quad 200 < q_u < 400$ kPa	150-300	24000-48000
$\quad q_u > 400$ kPa	> 300	> 48000

the manner of ASTM D 1195 [13]. An initial pre-pressure was applied to the stacked plates to represent the static weight of the foundation and the structure. Then, while maintaining the pre-pressure, the pressure level was fluctuated to represent the fluctuation of pressure during rotation of the radar antenna. This resulted in a hysteresis loop as shown for the term $k_{s\text{-}3}$ in Figure 8.10. It is not known how continuously repeatable this modulus was and whether the hysteresis was the result of compression and release of the air in the voids of the partially saturated soil being tested.

EXAMPLE PROBLEMS

EXAMPLE 8.1
A continuous wall footing 4 ft wide supports a load of 16 kips per foot (bearing pressure equals 4 ksf). The footing is assumed to be placed on or near the ground surface so that D_f equals 0. Within the significant depth below the footing, laboratory tests of soil samples indicate the soil weighs 120 lbs/ft³ and has R-test shear strength parameters of $c = 0.2$ ksf and $\phi_r = 32°$. What is the factor of safety of this foundation against bearing capacity failure?

Solution:
From Figure 8.4, $N_c = 35$, $N_q = 23$, and $N_\gamma = 22$.
From Equation 8.1:

$$q_d = cN_c + D_f N_q + 0.5\,\gamma B N_\gamma$$
$$q_d = 0.2 \times 35 + 0.12 \times 0 \times 23 + 0.5 \times 0.12 \times 5 \times 22 = 13.5 \text{ ksf}$$
$$FS = 13.5/4 = 3.4 > 3 \quad \text{OK}$$

EXAMPLE 8.2
A shallow rectangular footing will support a column load of 100 kips (445 kN). It is placed at a depth of 3 ft (0.9 m) below the surface of a sand layer 20 ft (6.1 m) thick. The sand has a moist unit weight of 125 pcf (2000 kg/m³) and the water table is at a depth of 12 ft (3.7 m). The average SPT, $N_{1(60)}$, for the several feet below footing level was 16 blows per foot. Limiting the settlement to 0.63 in. (16 mm) and a factor of safety of 3 against bearing capacity failure, what size footing should be used?

Solution:
First, solve for the footing size needed to limit settlement by using SI units and Equation 8.12 to limit settlement to 16 mm, $P/B^2 = 16Q + 0.67\,\sigma'$.

The effective preconsolidation pressure due to excavation for the footing 0.9 m is 18 kN/m². Because the width B is not known, it is necessary to solve Equation 8.12 for B.

$$\frac{445\text{kN}}{B^2} = 16\,\frac{16^{1.4}}{1.7B^{0.75}} + 0.67(18)$$

which reduces to $445 = 456.5B^{1.25} + 12B^2$

Solving by iteration, $B = 0.96$ m $= 3.15$ ft.

Use Equation 7.9 and Table 7.6 (b) to estimate the drained friction angle and then check the bearing capacity in English units.

Use Equation 8.3 and Figure 8.4 to estimate that $\phi' = 30 + 16/4 = 34°$. Then, $c = 0$, $N_q = 50$, and $N_\gamma = 35$, and

$$q_{ds} = (125 \times 3 \times 50 \times 0.4 \times 125 \times 3.15 \times 35) \div 1000 = 24.25 \text{ ksf}$$

The actual bearing pressure, $P/A = 80 / (3.15)^2 = 8.06$ ksf.
Therefore, the factor of safety against a bearing capacity failure $= 24.25/8.06 = 3.0$.

EXAMPLE 8.3

A shallow rectangular footing will support a column load of 100 kips (445 kN). It is placed at a depth of 3 ft (0.9 m) below the surface of a clay layer 30 ft (9.1 m) thick. The plasticity index of the clay is estimated, from field expedient tests, to be about 15. The clay has a moist unit weight of 120 pcf (1920 kg/m³) and the water table is at a depth of 12 ft (3.7 m). A field VST is used to estimate the undrained shear strength of the clay in a boring. The average undrained cohesion for the several feet below footing level is 3.0 kips/ft³. (a) For a factor of safety of 3 against bearing capacity failure, what size footing should be used? (b) Will the settlement be acceptable?

Solution:

a. The ultimate bearing capacity for the end of construction ($\phi = 0$) condition is given by Equation 8.6, for $q_{ult} = (3 \times 100) / B^2 = 300 / B^2$ ksf:

$$300 = B^2 [3.0 N_c + 0.12(3)] \text{ ksf}$$

from which the bearing capacity factor, N_c, can be determined as follows:

$$N_c = 6.2 + \frac{3/B}{0.45 + 0.24\,(3/B)}$$

Solving for B by iteration, $B = 3.63$ ft.

b. The estimated preconsolidation pressure is given by Equation 7.13,

$$p' = \frac{c_u}{0.11 + 0.0037 PI} = \frac{3.0}{0.11 + 0.0037(15)} = 18.1 \text{ ksf}$$

and the allowable bearing pressure is $P/A = 100 / (3.63)^2 = 7.6$ ksf. Because the bearing pressure is well below the preconsolidation pressure, the load will be in the recompression zone, and the settlement will be acceptably small.

EXAMPLE 8.4

Assume that the footing in Problem 8.2 is tilted at an angle of 20° from the horizontal. Calculate the maximum axial column load that can be supported with a factor of safety of 3.

Solution:
From Example Problem 2, $\phi' = 34°$, $B = 3.15$ ft, and $D/B = 3/3.15 = 0.95$. Interpolating in Figure 8.5 (b), $N_{\gamma q} = 80$. Then, using Equation 8.4 for a square footing (substituting $0.4B$ for $0.5B$), the ultimate pressure is

$$q_{ult} = cN_{cq} + 0.4\gamma BN_{\gamma q}$$

$$q_{ult} = 0 + 0.4\,(0.125)\,3.15\,(80) = 12.6 \text{ ksf}$$

For a factor of safety of 3, the allowable column load is $P = (12.6/3) \times 3.15^2 = 41.7$ kips.

EXAMPLE 8.5
Assume that the column footing in Problem 8.3 is located on a slope whose height $H = 20$ ft and whose slope angle $\beta = 15°$. From Problem 8.3, $D_f = 3$ ft, footing width $B = 3.63$ ft, $D/B = 0.83$, and cohesion $c = 3.0$ ksf. Then, from interpolation in Figure 8.7, Case II, $N_{cq} = 5.1$ and $N_{\gamma q} = 0$. Using Equation 8.4, the ultimate pressure is (see discussion above for Case I, step 7)

$$q_{ult} = cN_{cq} + \gamma DN_{\gamma q}$$

$$q_{ult} = 3.0\,(5.1) + 0.125\,(3.0)\,1 = 15.7 \text{ ksf}$$

For a factor of safety of 3, the allowable column load is $P = (15.7/3) \times 3.63^2 = 68.9$ kips.

EXAMPLE 8.6
After additional subsurface investigation, it is recognized that the undrained strength of the clay stratum in Problem 8.3 may be divided into two strata, the upper one having a thickness of 5 ft and a cohesion, c_1, varying uniformly from 4.0 ksf at the surface to 2.0 ksf at 5 ft. The lower clay stratum has a constant cohesion, c_2, of 2.0 ksf. For $B = 3.63$ ft, what is the allowable column load for a factor of safety of 3?

Solution:
From the statement of the problem, $c_2/c_1 = 2.0/4.0 = 0.5$; $B = 3.63$ ft; $T/B = (5-3)/3.63 = 0.55$; and $D/B = 3/3.63 = 0.83$. The equations and the graph in Figure 8.9 indicate that $N_c = 3.7$. The table on Figure 8.9 shows that for D/B 0.83, $N_{cd}/N_c = 1.22$. The equation at the bottom of Figure 8.9 for a rectangular footing indicates that $N_{cr} = N_{cd}$ (1.2). Therefore, $N_{cr} = 3.7 \times 1.22 \times 1.2 = 5.42$. The ultimate bearing capacity is:

$$q_{ult} = c_1 N_{cr} + \gamma D$$

$$q_{ult} = 4.0\,(5.42) + 0.125\,(3.0) = 22.0 \text{ ksf}$$

For a factor of safety of 3, the allowable column load is $P = (22.0/3) \times 3.63^2 = 96.6$ kips.

EXAMPLE 8.7
A plate load test is made on a clayey medium dense sand subgrade to provide soils data for pavement design. The load test is made on a stacked group of plates, the bottom plate of which is 30 inches in diameter, in accordance with ASTM D 1196 [13]. The load is applied in 1000-lb. increments and the deflection measured to the nearest 0.001 inches. The results of the plate load test are shown in Figure 8.11. Calculate the modulus of subgrade reaction. Is this a reasonable value for the soil tested? (Hint: check Table 8.2.)

Solution:
According to ASTM D 1196, the modulus of subgrade reaction is taken as the secant modulus at a deflection of 0.050 inches. The load-deflection curve of Figure 8.11 shows a deflection of 50/1000 of an inch at a plate load of 4650 lbs. The area of a 30-in. plate is 4.91 ft^2. The plate pressure is 4650 / 4.91 = 947 lb/ft^2 = 0.947 ksf. The coefficient of subgrade modulus is $k_s = q / \delta = 0.947 / (0.05/12) = 227$ kips/ft^3. Table 8.2 indicates that the subgrade modulus for a clayey medium-dense sand typically ranges from 200 to 500 kcf. The test result is consistent with normal values for this type of soil.

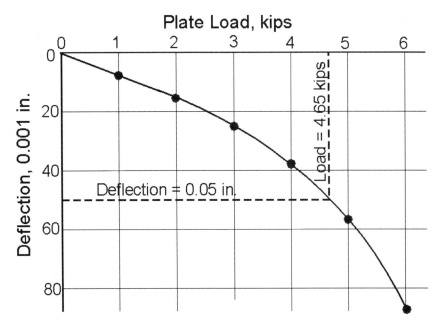

FIGURE 8.11 Load-deformation graph to accompany Problem 8.7.

REFERENCES

1. Spigolon, S. J. 1963. "Investigation of Size Effect for Circular Footings on Loess." Doctoral dissertation. University of Illinois, Champaign-Urbana.
2. Terzaghi, K. 1943. *Theoretical Soil Mechanics*. New York: John Wiley & Sons.
3. Meyerhof, G. G. 1955. "Influence of Roughness of Base and Ground-water Conditions on the Ultimate Bearing Capacity of Foundations." *Géotechnique*. Vol. 5, pp 227-242.
4. Terzaghi, K., Peck, R. B., and Mesri, G. 1996. *Soil Mechanics in Engineering Practice*, 3rd ed., p.412. New York: John Wiley & Sons.
5. Meyerhof, G. G. 1953. "The Bearing Capacity of Foundations under Eccentric and Inclined Loads." *Proceedings of the Third International Conference on Soil Mechanics and Foundation Engineering*. Vol. 1, pp. 440-445. Zurich.
6. NAVFAC DM-7.02. 1982. *Foundations and Earthwork, Design Manual 7.02*, U. S. Department of the Navy, Naval Facilities Engineering Command. Alexandria, VA. September.
7. Meyerhof, G. G. 1957. "The Ultimate Bearing Capacity of Foundations on Slopes." *Proceedings of the Fourth International Conference on Soil Mechanics and Foundation Engineering*. London.
8. Skempton, A. W. 1948. "The $\phi = 0$ Analysis of Stability and its Theoretical Basis." *Proceedings of the Second International Conference on Soil Mechanics and Foundation Engineering*, Vol. 1, pp. 72-78. Rotterdam.
9. Button, S. J. 1953. "The Bearing Capacity of Footings on a Two-layer Cohesive Subsoil," *Proceedings, Third International Conference on Soil Mechanics and Foundation Engineering*, Vol. 1, p 332. Zurich.
10. Sowers, G. F. 1979. *Introductory Soil Mechanics and Foundations: Geotechnical Engineering*, Fourth Edition. New York: Macmillan.
11. Bowles, J. E. 1988. *Foundation Analysis and Design*, Fourth Edition. New York: McGraw-Hill.
12. NAVFAC DM-7.01. *Soil Mechanics, Design Manual 7.01*, U.S. Department of the Navy, Naval Facilities Engineering Command, Alexandria, VA. September.
13. ASTM. "Natural Building Stones; Soil and Rock; Geotextiles," *1999 Annual Book of ASTM Standards*, Vol. 04.08. American Society for Testing and Materials. Philadelphia, PA.
14. Weissmann, G. F. Feb. 1965. "Measuring the Modulus of Subgrade Reaction," *Materials Research and Standards*. American Society for Testing and Materials, Philadelphia, PA.

CHAPTER 9
DEEP FOUNDATIONS—PILES AND PIERS

Piles and drilled shafts (also called piers) are used to transfer a surface load to a competent soil or rock layer at depth when the surface layer is not adequate or is not as economically feasible to use. This chapter deals with those foundation elements that are *long* and *slender*. They are differentiated from *footing* foundations in that the ratio of the depth of the foundation to the tip or base diameter or width usually is greater than 4, whereas for footings this ratio is commonly less than unity [1]. Deep foundations can be driven, drilled, or dug.

TYPES OF DEEP FOUNDATIONS

The distinction between piles and piers is somewhat arbitrary. Piles are usually small in diameter, are typically installed by driving with a hammer or vibrator, and often are grouped into clusters or rows for supporting column loads. Piers are usually a larger cross section and typically are constructed to support a single column. A pile, however, also can be constructed by predrilling a hole and then filling it with concrete and reinforcement, a process similar to that for construction of a drilled shaft, or pier. Typical pile characteristics are described in Table 9.1.

Piers fall into two categories: an underground structural member that transmits a column load to a competent stratum at depth, and a support of concrete or masonry for the superstructure of a bridge, most often extending through a body of water to a level above the maximum high water [1]. Only the first category will be considered below.

Piers, or drilled shafts, generally are constructed by machine-drilling a large-diameter hole and then filling it with concrete. Some very large-diameter shafts in the past were hand-excavated. The concrete can be reinforced and the bottom of the shaft can be enlarged, or belled, to increase the bearing area.

TABLE 9.1 Typical Characteristics of Piles [2]

Pile Type	Typical Length Feet	Typical Design Load, Tons	Comments
Timber	30-60	10-50	Friction pile. Easy to handle. Becoming expensive. Requires treatment.
Steel-H section	40-100	40-120	High capacity, low displacement. Will penetrate small obstructions. Corrodible. Spliceable. End bearing on rock.
Precast concrete Prestressed concrete	40-50 60-100	Wide Range	High load capacities. Corrosion resistant. Hard driving possible. Not spliceable.
Mandrel-driven thin-shell Cast-in-place concrete	50-80	Wide Range	Friction piles in granular materials. Thin shell driven with mandrel, then concrete filled. Initial economy. Shell easily damaged. Difficult to splice.
Thick-shell (no mandrel)	30-80	50-70	Friction piles of medium length. Tough shell. Can be re-driven. Difficult to splice.
Pressure-injected footings	10-60	60-120	Best in granular soils. Hole made by drill or thin shell. Concrete rammed to make enlarged bulbous base.
Steel pipe piles (with concrete core)	40-120+	80-120 500-1500	High load and high bending resistance. Moderate cost in some areas. Easy to splice. Can be cleaned out if open ended.
Auger-placed, pressure-injected cement grout	30-100+	35-70+	Friction or end bearing. Economical. Non-displacement. Quality depends on continuous grout placement.

DEEP FOUNDATION FAILURE MODES

Figure 9.1 shows the five conventional modes (a, b, c, d and e) by which piles and drilled shafts fail under axial loading. All axial loads are resisted by the end-bearing capacity and by the skin friction on the shaft. Piles normally are used in groups although analysis of pile capacity usually is done for a single pile. Piers rarely are used in groups.

Figure 9.1(a) indicates a pile group and a single pile or pier, each supporting the load P in end-bearing on a rock or hard stratum. Although skin friction with the softer soil is present, this form of support is usually so small compared to the end bearing that it normally is ignored in design calculations.

In Figure 9.1(b), the pile or pier embedded in a firm layer is acted on by a clay soil that swells after pile placement. The clay pulls upward on the perimeter of the pile or pier, reducing the end-bearing and causing a reversal of skin friction direction.

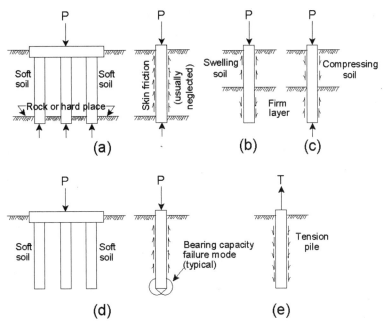

FIGURE 9.1 Typical loading configurations for piles or piers under axial loading.

When an end-bearing pile or pier is placed through soil that is likely to compress, as in Figure 9.1(c), typically because of additional fill placed after the pile or pier is placed, the consolidation of the soft layer under the new loading creates an additional downward force on the perimeter of the pile or pier. This additional downward force increases skin friction and end bearing on the section of pile or pier in the firm layer.

Figure 9.1(d) shows a pile group and a single pile completely supported by skin friction. Piers are not normally used in this situation. The end-bearing is normally so small, because of the small end area of the pile, that it normally is ignored in design calculations. A typical end-bearing shear-failure pattern for friction piles is shown for the single pile. A similar pattern holds for piles in groups if they are located far enough apart to permit full development of the shear pattern.

Figure 9.1(e) displays a pile or pier acted on by an uplift, or tension, load. In this situation, the only resistance to the tension force is provided by skin friction.

DEEP FOUNDATION CAPACITY FROM LOAD TESTS

Three methods commonly are used to determine the ultimate load capacity of a pile. In order of decreasing reliability, and decreasing cost, they are the full-scale load test, the static analysis, and the pile-driving formula. Of these, only the first two are applicable for a pier.

The most reliable and most costly means of evaluating the load capacity of a deep foundation is by means of a full-size load test. Load tests can be made during the design phase or during construction to verify, or proof, design loads.

Load test methods for piles have been standardized as ASTM D 1143 [3] for "Piles under Static Axial Compressive Load," as ASTM D 3689 for "Individual Piles under Static Axial Tensile Load," and as ASTM D 3966 for "Piles under Lateral Loads." Load tests of drilled shafts are not standardized, but generally are adapted from a combination of a pile load test method and the plate load test method described in Chapter 8. All axial compression load tests require a reaction load of at least twice the design load, using either a dead load or a beam connected to a group of tension piles.

Field Loading Tests in Axial Compression

The three standardized field pile load test procedures for axial compression tests described in ASTM D 1143 are described below.

Standard Loading Procedure. This test also is known as the *slow maintained-load method*. An individual pile is loaded to 200 percent of design load, or to 150 percent of group design load, in increments of 25 percent of design load. Each increment is held until the rate of settlement is less than 0.01 in. (0.25 mm), but not for more than 2 hours. It is normal to include an unloading-reloading cycle, after 100 percent of the design load is reached, to establish the elastic rebound of the pile. The final load at 200 percent of design load is maintained for 24 hours, after which the pile is unloaded in increments. This test can take two to three days to complete.

Peck, Hanson, and Thornburn [1] consider that the cost of maintaining the final load for 24 or more hours is rarely justified. They suggest that, instead "[i]n most instances the performance under twice the design load in 24 hr or more can be judged by applying 2.25 times the design load for 1 hr."

Quick Load Test Method for Individual Piles. This short duration test also is known as the *quick maintained-load method*. The load is applied in increments of from 10 percent to 15 percent of the design load with a constant time interval of 2.5 minutes between load increments. The maximum specified load is held for five minutes and the load is removed in increments in the same manner as for loading. NAVFAC DM 7.02 [2] indicates this test generally is loaded to 300 percent of design load.

Constant Rate of Penetration Test for Individual Piles. This is a short duration test using displacement control. The applied load is varied as needed to maintain a pile penetration rate of from 0.01 to 0.05 in. (.25 to 1.25 mm) per minute for cohesive soil or from 0.03 to 0.10 in. (0.75 to 2.5 mm) per minute for granular soils. Loading is continued until no further load increase is needed to maintain the penetration rate, essentially until a plunging failure is reached. This method is particularly suited to testing piles in cohesive soils or where only the ultimate capacity is to be measured.

Tension and Lateral Loading Tests

Pile load test procedures for tension piles, in ASTM D 3689 [3], are essentially the same as those for axial compression. Lateral load testing, according to ASTM D 3966 [3], includes a standard loading sequence as in axial loading except that the loads are applied in increments of 25 percent of design load and each increment is held for 10 to 20 minutes. The 200 percent of design load is held for 60 minutes, after which the pile is unloaded in increments. The standard test method also provides for surge loading in which a specified number of loading-unloading cycles are applied.

Test Loading and Time Relationship

The physical action of driving a displacement pile into the ground causes a disturbance of the surrounding soil to two or more radii away from the pile surface, and causes excess pore water pressures to be generated. If tested immediately after placement, the pile bearing capacity will be greatly underestimated. With a sufficient time lapse between driving and testing, the excess pore water pressures are dissipated and the surrounding soil regains strength. This often is called *freezing*.

Piles in granular soils should be tested no more than 24 to 48 hours after driving to allow time for excess pore water pressures to be dissipated. A substantial increase in load bearing capacity can occur if the load tests are made two or more weeks after driving. Piles in cohesive soils generally are tested 30 to 90 days after driving; because of the much lower permeability of clayey soils, more time is needed for the excess pore water pressure to dissipate. For cast-in-place concrete piles, the time needed for development of concrete strength is also a time-delay determinant.

Interpretation of Pile Load Tests

Typical pile load test results are shown in Figure 9.2. Friction piles in cohesive soils will, at ultimate load, reach a plunging failure (curve *a*) at which a large settlement will occur with no further increase in load. A plunging failure of this type can be observed only if the pile test is carried to a sufficiently high load. An end-bearing pile (curve *b*) and/or a pile deriving support from both end bearing and skin friction, however, will reach the plunging failure mode rarely, if ever.

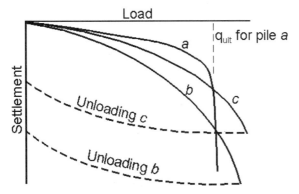

FIGURE 9.2 Typical results of loading tests on (a) friction pile, (b) end-bearing pile or pier, and (c) pile having combined end bearing and skin friction support (adapted from [1]).

Figure 9.2 also shows unloading, or rebound, curves b and c, which are assumed to be due to the recoverable elastic component of the load-settlement curve. By subtracting the *elastic* rebound from all measured settlements, it is possible to plot a *net* settlement curve, which is assumed to be due only to permanent settlement.

It is readily seen in Figure 9.2 that for curves b and c the determination of an ultimate capacity load is very difficult to establish rationally for tests that do not result in a plunging failure. A number of methods have been proposed in various building codes for establishing the failure load unambiguously. These include such diverse empirical methods as (a) finding the load that occurs at one-half the settlement where there is a sharp break in the load-settlement curve in the net settlement curve, (b) finding the load that occurs at three-quarters of the net settlement at twice the design load, and (c) finding the load that occurs at some other fraction of the net settlement at twice, or some other factor, of the design load.

A reasonable procedure is one developed by Davisson [4] and illustrated in Figure 9.3. The procedure assumes that the pile is a free column having an elastic compression equal to

$$\delta_e = \frac{PL}{AE} \tag{9.1}$$

where δ_e = Elastic compression
 P = Axial load on pile
 L = Pile length (for end-bearing pile)
 A = Cross-sectional area of pile material
 E = Modulus of elasticity of pile material.

Typical values for the modulus of elasticity for use in Equation 9.1 include (a) for steel piles or mandrels, 29,000,000 psi; (b) for concrete piles, 3,000,000 ± psi; and (c) for timber piles, from 800,000 to 1,600,000 psi (use 1,000,000 psi as an average). An alternative to calculating Equation 9.1 is using the elastic rebound from the load-settlement curve, as shown in Figure 9.2.

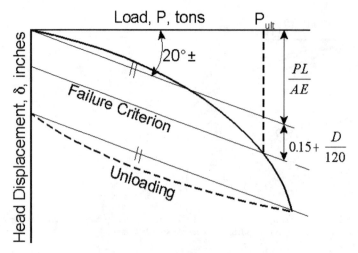

FIGURE 9.3 Davisson [4] criterion for determining failure load in a pile load test.

The Davisson procedure calls for plotting the gross load-settlement curve at such a scale that the straight line of the elastic compression from Equation 9.1 will be at an angle of about 20° as shown. A parallel line is drawn at a pile head displacement of

$$S_f = \delta_e + \left(0.15 + \frac{D}{120}\right) \quad (9.2)$$

where S_f = Displacement at failure in inches
 δ_e = Elastic compression (Equation 9.1)
 D = Pile diameter in inches.

The intersection of the line with the load-settlement curve is considered the failure load and the usual allowable pile load is one-half of that value. In practice, it is common to use the unloading, or rebound, curve to establish the elastic compression part of the displacement. If rebounding is done at an intermediate load and at the 200 percent load, an average value is used. If the observed load-settlement curve does not cross the failure criterion line, the maximum test load is taken as the failure load.

This procedure is based on the amount of pile-tip displacement required to develop full bearing capacity. Peck, Hanson, and Thornburn [1] indicate this criterion is applicable only to tests in which the load increments are held for one hour or less, as in the abbreviated standard load test or the quick load test described above. If the load at 200 percent of design is maintained for 24 hours or longer, the application of the Davisson criterion is overly conservative due to creep or consolidation settlement.

DEEP FOUNDATION CAPACITY FROM STATIC ANALYSIS

The load-carrying capacity of a single isolated pile or pier is derived from the bearing capacity at the end (pile tip) and from the frictional resistance of the soil developed around the shaft, as defined in Equation 9.3,

$$Q_{ult} = Q_b + Q_s - W_p \quad (9.3)$$

where Q_{ult} = Ultimate vertical load capacity of pile or pier
 Q_b = Component of load capacity due to bearing capacity at pile or pier base
 Q_s = Component of load capacity due to side friction
 W_p = Net weight of pile or pier, above weight of displaced soil.

The load capacity generated at the tip is the area of the base times the unit-bearing capacity, as given in Equation 9.4,

$$Q_b = A_b\left(cN_c + \sigma_t' N_q + 0.5\, B\gamma_b' N_\gamma\right) \quad (9.4)$$

where A_b = Area of pile or pier base
 c = Soil cohesion
 σ_t = Effective vertical stress at pile or pier base
 B = Base diameter
 γ_b' = Effective unit weight of soil in the failure zone beneath base
 N_c, N_q, N_γ = Bearing capacity factors.

The load capacity due to skin friction on the shaft of the pile is given in Equation 9.5,

$$Q_s = \int_0^L f_s C_z \, dz \tag{9.5}$$

where f_s = Ultimate skin resistance per unit area of pile shaft at depth z
C_z = Effective perimeter of pile at depth z
L = Length of pile in contact with soil
dz = Increment of length at depth z.

Load Carrying Capacity of a Single Pile or Pier in Granular Soil

Pile or Pier in Compression. For deep foundations without enlarged tips, Equations 9.4 and 9.5 can be simplified and combined for granular soils, in which $c = 0$, corresponding to the parameters shown in Figure 9.4. This can be done by eliminating the N_γ term,

$$Q_{ult} = \sigma_t' A_b N_q + \sum_H^{L=H+D} \sigma_o' K_{hc} \tan \delta P \Delta L \tag{9.6}$$

where N_q = Bearing capacity factor (see Table 9.2)
σ_o' = Effective overburden pressure at depth L_o
K_{hc} = Ratio of horizontal to vertical pressure–pile in compression (see Table 9.3)
δ = friction angle between pile and soil (see Table 9.4)
P = Perimeter or circumference of pile at depth L_o
L_o = Mid-depth of increment over which lateral pressure is applied
ΔL = Increment of length over which lateral pressure is applied.

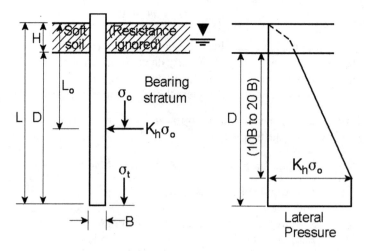

FIGURE 9.4 Load carrying capacity of single pile or pier in granular soil [2].

TABLE 9.2 Bearing Capacity Factors, N_q, For Single Pile in Granular Soil [2, p. 194]

ϕ', Degrees	N_q, Driven Pile (Displacement)	N_q, Drilled Piles or Piers†
26	10	5
28	15	8
30	21	10
31	24	12
32	29	14
33	35	17
34	42	21
35	50	25
36	62	30
37	77	38
38	86	43
39	120	60
40	145	72

* Limit ϕ to 28° if jetting used.

† If a bailer or grab bucket is used below groundwater table, calculate end bearing based on ϕ not exceeding 28°.
For piers with greater than 24-in. diameter, settlement rather than bearing capacity usually governs design. For estimating settlement, use 50% of the settlement for an equivalent footing resting on the surface of comparable granular soils

TABLE 9.3 Horizontal Earth-Pressure Coefficients for Single Pile in Granular Soil [2]

Pile Type	Compression, K_{hc}	Tension, K_{ht}
Driven, single H-pile	0.5-1.0	0.3-0.5
Driven, single displacement pile	1.0-1.5	0.6-1.0
Driven, single displacement tapered pile	1.5-2.0	1.0-1.3
Driven jetted pile	0.4-0.9	0.3-0.6
Drilled pile, less than 24-in. diameter	0.7	0.4

TABLE 9.4 Friction Angle, δ For Single Pile in Granular Soil

Pile Type	Friction angle, δ [2, p. 194]	Friction angle, δ [5, p. 18]
Concrete	0.75 ϕ	0.90 ϕ to 1.0 ϕ
Steel	20°	0.67 ϕ to 0.83ϕ
Timber	0.75 ϕ	0.80 ϕ to 1.0ϕ

Experimental and field evidence [2] [5] [6] indicate that both bearing capacity and skin friction increase with vertical effective stress σ_o up to a limiting depth of embedment. The limiting depth depends on the position of the water table and the relative density of the cohesionless soil. Beyond a limiting depth of about 10 to 20 pile diameters, there is only an insignificant increase in end-bearing capacity and the increase in side friction is directly proportional to the surface area of the pile. For design purposes, the critical depth is assumed as

D_c = 10B for loose sands or silts
D_c = 15B for medium dense sands or silts
D_c = 20B for dense sands or silts.

Therefore, if depth D is greater than from 10B to 20B, depending on the relative density of the sand, limit the vertical effective stress σ_o to that value corresponding to D = 10B to 20B. This is shown graphically at the right side of Figure 9.4.

Step-taper and Tapered Piles. Step-taper and tapered piles depend on the decreasing pile diameter to increase skin friction. Terzaghi, Peck, and Mesri [6], based on the work of Norldund [7], indicate that the effect of the tapering can be accounted for by replacing $\tan \delta$ in Equation 9.6 with $\sin(\omega + \delta) \sec \omega$. The pile perimeter P in Equation 9.6 must be adjusted for each increment of length used in the calculation.

Tensile (Pullout) Loading of Piles and Piers. A pile or pier in tension reverses the direction of the skin friction as shown in Figure 9.1(e). The end bearing is, of course, no longer a factor. Therefore, for a pile or pier in tension, only the skin friction component of Equation 9.6 is effective, using the horizontal earth-pressure coefficients for tension, K_{ht}, in Table 9.3.

Load Carrying Capacity of a Single Pile or Pier in Cohesive Soil

For a deep foundation in saturated cohesive soil subjected to a rapidly applied (end of construction) load, the undrained soil strength parameters are $\phi = 0°$ and $c = 0.5 q_u$, where q_u is the unconfined compressive strength, and $N_q = 0$ (Figure 8.4). Therefore, as shown in Figure 9.5, the ultimate bearing capacity of a pile or pier base in clay from Equation 9.4 becomes that in Equation 9.7,

$$Q_{b-ult} = A_b c N_c F_r \qquad (9.7)$$

where $c N_c F_r < 80$ ksf (40 tsf, 4000 kPa).

The bearing capacity factor N_c usually used in design is that defined in Equation 8.9 and has an upper limit of nine for foundations with $D_f/B > 4$. In cases where the clay at the base might have been softened, c can be reduced by one-third and could cause a local bearing capacity failure. The factor F_r should be 1.0 except when the base diameter B exceeds about 6 feet (1.8 m) for which [7]

$$F_r = \frac{2.5}{aB + 2.5_b}, F_r \leq 1.0 \qquad (9.8)$$

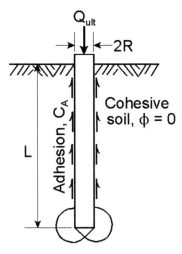

FIGURE 9.5 Load carrying capacity of single pile or pier in cohesive soil [2].

where $a = 0.0852 + 0.0252\,(L/B)$, $a \leq 0.18$
$b = 0.45\,c$, $0.5 \leq b \leq 1.5$

Skin Friction for Driven Piles. Two unique factors affect the load bearing capacity of a driven pile in cohesive soil: (a) the vertical movement needed to develop tip resistance of piles may be several times larger than that required to develop the skin friction resistance, and (b) the physical act of driving a pile tends to cause disturbance of the surrounding soil, a so-called *socket* effect. The first of the two effects is generally dealt with by eliminating end-bearing from consideration in determining the load carrying capacity and using only skin friction.

The second factor involves using adhesion c_a instead of the undrained cohesion c_u. This factor varies considerably, being affected by pile type, soil type, and the method of installation. Defining the adhesion factor α as the ratio of c_a and c_u, a commonly used relationship [5], is

$$\alpha = c_a / c_u = 1.0, \quad c_u \leq 0.25 \text{ tsf} \tag{9.9a}$$

$$\alpha = c_a / c_u = 1.25 - c_u, \quad 0.25 < c_u < 0.75 \text{ tsf} \tag{9.9b}$$

$$\alpha = c_a / c_u = 0.5, \quad c_u > 0.75 \text{ tsf} \tag{9.9c}$$

Therefore, modifying Equation 9.5 to account for the adhesion factors from Equation 9.9, the ultimate skin friction on single piles or piers in cohesive soils, in both compression and tension, is

$$Q_{s-ult} = \sum_0^L \alpha\, c_u\, P_z\, \Delta z \tag{9.10}$$

Skin Friction on Drilled Shafts. The skin friction on a drilled shaft varies along the length of the pile or pier, increasing with depth to a maximum and then decreasing toward the tip or base. A suggested design procedure [7] states that for straight shafts, the skin friction is ignored for the uppermost and lowermost 5 ft (1.5 m) and the adhesion factor $\alpha = 0.55$ at all other points, with a maximum skin friction of 2.75 tsf (275 kPa).

Lateral Loading of Piles and Piers

A pile or pier that is loaded by moment or lateral thrust at its top resists the load by deflecting sufficiently to mobilize the lateral resistance of the surrounding soil. The amount and distribution of the lateral resistences are a function of the relative stiffnesses of both the pile or pier and the soil. Because soil has a non-linear stress-strain relationship (Chapter 8), solutions for lateral loading are complex.

Theoretical equations for determining the ultimate lateral load and/or the ultimate lateral deflection at the ground line, contained in Engineer Manual EM 1110-1-1905 [9], can give reasonably good results and are recommended for an initial estimate. The COM624P computer program developed by Wang and Reese, described in [10], is widely used. These methods are valid only for initial estimates because they require that the soil stress-strain properties be well known. The actual lateral load capacity ultimately should be determined or verified by a field load test.

Bearing Capacity of Pile Groups

The term *group efficiency* is the ratio of the bearing capacity of a pile group to the sum of the bearing capacities of the single piles in the group. The bearing capacity of a group of piles is generally less than the sum of the capacities of the individual piles [2].

Group Capacity in Rock. If the individual piles in a group are spaced normally and are installed to rock, the group efficiency $G_e = 1.0$. Failure of the group is a consideration only if the rock surface is sloping or if sliding can occur along unfavorable dipping, weak, planes.

Group Capacity in Soil. Each pile in a pile group founded in soil acts like an individual pile if the individual piles are spaced at least seven pile diameters apart. At closer spacing, a pile group in soil acts as a large pier if the piles and the confined soil sink as a unit. In this mode, it is necessary to consider the stress distributed to a lower stratum (Chapter 4) and the effect this can have on the bearing capacity of the lower soil layers.

The ultimate bearing capacity of a pile group in granular soil is calculated using Equation 9.6. The ultimate bearing capacity of a pile group in cohesive soil is calculated using the combination of Equations 9.7, 9.8, and 9.10. In both cases, (a) the base area A_b is the area bounded by the perimeter at the base of the group and (b) the perimeter P is the outer perimeter of the pile group at any given depth.

Settlement of Pile Groups

Pile Group in Granular Soil. NAVFAC DM 7.02 [2] presents the following method for determining the settlement of a pile group in granular soil, assuming there are no soft, compressible layers within a significant depth below the pile tips:

$$S_g = S_i \sqrt{B/D} \qquad (9.11)$$

where S_g = Settlement of pile group
S_i = Settlement of a single pile estimated or determined from load tests
B = Smallest dimension of pile group
D = Diameter of individual pile.

In the case where a compressible stratum exists below the pile group, and the group is essentially end-bearing, then treat the group as a pier and estimate the pressure distribution as shown in Figure 4.16, using the 2:1 rule, starting from the base of the group. If the group consists essentially of friction piles, use the method described below for cohesive soils. Calculate the compression settlement of the lower layer, as discussed in Chapter 4 for cohesive soils or in Chapter 8 for granular soils.

Pile Group in Cohesive Soil. To determine the pressure on a lower stratum due to a friction pile group, assume the base of the group starts at a point two-thirds of the length of the piles from the ground surface. Then, use the 2:1 rule of Figure 4.16, as shown in Figure 9.6. Calculate the compression settlement of the lower layer, as discussed in Chapter 4 for cohesive soils or in Chapter 8 for granular soils.

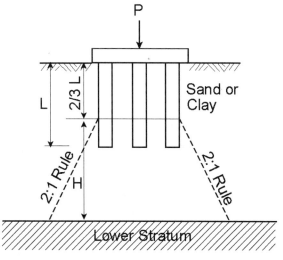

FIGURE 9.6 Distribution of pressure from pile group onto lower stratum.

PILE CAPACITY FROM DRIVING DATA

The use of a pile-driving formula derives from the observation that the bearing capacity of a pile driven to end-bearing seems to increase with the resistance to driving. Although this general concept appears to be valid, all attempts to develop a relationship between bearing capacity and the driving resistance, blows of the hammer per unit of penetration, have been unsuccessful. One important factor is that the pile resistance during driving does not take into account the strength increase with time due to pore water pressure dissipation (*freeze*). The application of any formula to friction piles, for which the underlying theory does not apply, has been shown [6] to be completely invalid.

Engineering News Formula

The formerly widely used pile-driving formula known as the *Engineering News* formula was developed in the late 1800s for timber piles driven by a drop hammer. This formula has also been thoroughly discredited [6] because statistical studies have shown such poor correlation and such poor precision (wide scatter). Ttherefore, its use for any purpose should be discouraged.

Danish Formula

One of the few formulae considered to have a reasonable precision, based on a statistical study of pile load tests, is the *Danish* formula [11], which should be used with a factor of safety of 3,

$$Q_{dy} = \frac{\alpha W_H H}{S + 0.5 S_e} \quad (9.12)$$

in which

$$S_e = \sqrt{\frac{2\alpha W_H H L}{AE}} \quad (9.13)$$

where Q_{dy} = Ultimate dynamic bearing capacity of driven pile
α = Pile driving hammer efficiency (normally 1)
W_H = Weight of hammer
H = Hammer drop (note that $W_H H$ = Hammer energy)
S = Inelastic set of pile, in distance per hammer blow
S_e = Elastic set of pile, in distance per hammer blow
L = Pile length
A = Pile end area
E = Modulus of elasticity of pile material (see discussion following Equation 9.1).

Use of Pile Driving Formulae

A pile-driving formula should not be used alone to establish the bearing capacity of a single pile, except on those very small structures for which the geology is well known, there is extensive previous experience to be used as a guide, and a suitable factor of safety is applied.

In practice, a pile driving formula such as the Danish Formula can be used as a first approximation for establishing the pile placement depth for a field load test. The pile placement depth should also be checked with a static analysis. Ideally, the test location should be at, or next to, that of an SPT or VST boring or of a CPT sounding. Then, if the pile load test is satisfactory, and if a study of the local geology combined with the subsurface investigation indicate a reasonably uniform site, all other similar piles can be driven to the same driving resistance, in hammer blows per unit of pile penetration.

EXAMPLE PROBLEMS

EXAMPLE 9.1
An industrial structure with high column loads is to be constructed on very soft alluvial flood plain deposits along the Mississippi River. A review of geologic literature indicates there is a layer of hard clay at a depth of about 100 feet. This is substantiated by a subsurface investigation of the site. A pile foundation is recommended. Which types of piles should be considered (from Table 9.1) and why?

Solution:
Of the pile types listed in Table 9.1: (a) steel H-section can be used, but the end must be covered with a shoe to prevent cutting into the hard clay; (b) prestressed concrete piles can be crane-handled in 100-ft lengths; it may be necessary to cut off the upper part of the pile if the lower stratum is slightly uneven; (c) steel pipe piles can be used, with sections weld-spliced if needed; pile capacity can be increased by filling with concrete; (d) pressure-grouted (auger cast) piles can be (and have been) used; they require close quality control to ensure no void or *necked down* areas occur. No other pile types are suitable. For example, timber piles are difficult to find in 100-ft lengths and can be very expensive.

EXAMPLE 9.2
A pile load test has been made on a closed-end steel pipe pile, 50 feet long, 12 inch OD, with ¼-inch wall thickness, driven through a soft clay to end-bearing in a dense sand. The design capacity of the pile is 50 tons. The load test procedure was the standard loading method, except that loading to 230 percent of design load was used instead of holding the final 200 percent load for 24 hours or more. The plot of load versus settlement of the pile top is shown in Figure 9.7. Based on the criterion shown in Figure 9.3, determine the ultimate load capacity of this pile. Is it satisfactory for the design load?

Solution:
The pile has a cross-sectional area of steel of 9.23 square inches. The elastic compression of the steel pile is determined by Equation 9.1 and, at $P = 100$ tons and $E = 29,000,000$ psi, is equal to 0.44 inches. The field-test data plot is scaled so that the initial part of the curve is about 20° from the horizontal. The permanent set criterion, Equation 9.2, is calculated using $B = 12$ inches, yielding 0.25 inches. This offset line is drawn parallel to the elastic compression line. The failure criterion line crosses the load-settlement curve at the failure load, in this case 104 tons. As a check on the elastic line, the rebound curve is approximated with a straight line and compared. In this test, they are reasonably similar.

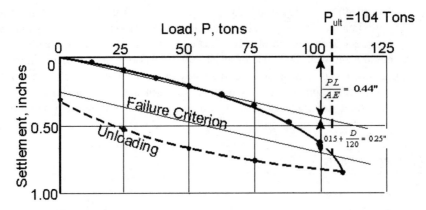

FIGURE 9.7 Results of field load test of pipe pile.

The ultimate load capacity of 104 tons is 2.1 times the design load. This is slightly more than the normal factor of safety of 2 for piles when tested by a field load test. Therefore, the pile is acceptable for the design load.

EXAMPLE 9.3
A precast concrete pile, having a diameter of 16 inches, is driven through a 15 ft thick soft clay layer into a medium-dense sand deposit. The sand has a uniform drained friction angle of 34°. The water table is at the bottom of the clay layer. The unit weight of the clay is 115 lb/ft³ and the saturated unit weight of the sand is 125 lb/ft³. Neglecting the skin friction in the clay layer, how far into the sand layer must the pile be driven to develop an allowable load capacity of 50 tons with a factor of safety of 3?

Solution:
Refer to Figure 9.8, which is similar to Figure 9.4.
   For a factor of safety of 3, the required ultimate load calculated from static equations is
      3 (50) = 150 tons.
   The critical depth for a medium dense sand is 15 B = 17.5 ft.
   For ϕ = 34°, N_q = 42 from Table 9.2.
   For a driven displacement pile in compression, K_{hc} = 1.0 to 1.5 (Table 9.3); use 1.25.
   For a concrete pile, δ = 3/4 ϕ = 0.75 (34°) = 25.5°
   Pile area = 1.40 ft² and pile circumference = 4.19 ft.
   For the pile tip at 17.5 ft below the clay layer:
   Effective vertical pressure = 15 (115) + 17.5 (125 – 62.4) = 2820 psf
   End bearing Q_b = 2820 (1.40) 42/2000 = 77.5 tons (Equation 9.6)
   Skin friction Q_s = 1.25 (1725 + 2820)/2 (4.19) tan 25.5° (17.5) = 52 tons (Equation 9.6)
   Total calculated bearing capacity at D = 17.5 ft = 77.5 + 52 = 129.5 tons < 150 tons.
   The pile must be deepened to develop additional 20.5 tons skin friction.
   The effective vertical stress below 17.5 remains constant.
   Skin friction Q_s = 20.5 tons = 1.25 (2820) 4.19 (tan 25.5°) ΔL = 3.5 ΔL.
   Solving, ΔL = 5.9 ft.
   Therefore, the pile must extend 5.9 + 17.5 = 23.4 ft below the bottom of the clay layer.

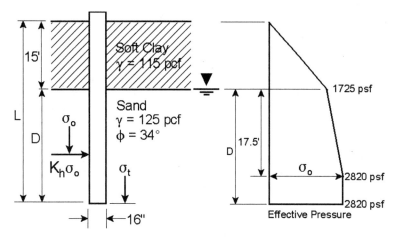

FIGURE 9.8 Loading conditions for 16-inch precast concrete pile.

EXAMPLE 9.4
What is the allowable uplift capacity of the pile analyzed in Problem 9.3?

Solution:
The pile will develop skin friction in the sand over a length of 23.4 ft.
 For a driven displacement pile in tension, K_{ht} = 0.6 to 1.0 (Table 9.3); use 0.80.
 Using skin friction only, Q_s = 0.80 (1725 + 2820)/2 (4.19) tan 25.5° (17.5) + 0.80 (2820) 4.19 (tan 25.5°) 5.9 = 31.8 + 13.3 = 45.1 tons.
 The allowable uplift is $Q_s/3$ = 15 tons plus the weight of the pile.
 The volume of the pile is 1.40 (23.4 + 15) = 53.75 ft³. Concrete has a unit weight of 150 pcf, so the total weight of the pile is 4.0 tons.
 Therefore, the allowable uplift load is 15 + 4 = 19 tons.

EXAMPLE 9.5
A straight-sided, drilled concrete pier, 10 ft in diameter, is used to transfer a structure column load to a stiff clay layer at a depth of 25 ft below the surface. The clay has a uniform unconfined compressive strength of 1.5 tons/ft² from the surface to at least 15 feet below the pier base. There are no softer layers below that level. What is the allowable column load on the pier, using a factor of safety of 3?

Solution:
For unconfined compressive strength equal to 1.5 tsf, the cohesion c = 0.75 tons/ft².
 The base area of the 10 ft diameter drilled shaft is 78.54 ft².
 The ultimate bearing capacity of the pier is given by Equation 9.7: Q_{ult} = 78.54 (0.75) 9 (F_r).
 Because the pier width is greater than 6 ft, F_r is less than 1.0. Then, Equation 9.8 states that

$$F_r = \frac{2.5}{aB + 2.5b}$$

where $a = 0.0852 + 0.0252\,(25/10) = 0.148$
$b = 0.45\,(1.5) = 0.675$
$F_r = 0.79$.

Skin friction on a drilled pier is limited to only the central part of the depth; the uppermost and lowermost 5 ft (1.5 m) are not effective. The adhesion factor $\alpha = 0.55$ for piers. Then, $Q_{ult} = Q_b + Q_s = 78.54\,(0.75)\,9\,(0.79) + 15\,(0.55)\,0.75\,(31.4) = 613$ tons and the allowable pier load $Q_{allow} = 613\,/\,3 = 204$ tons.

EXAMPLE 9.6
Several 18-inch diameter pressure-grouted piles are to be augered into the clay layer of Problem 9.5 instead of the pier. Each pile is far enough away from the other piles to act individually. Assuming the unconfined compressive strength of the clay layer is constant and is equal to 1.5 tsf, to what depth must each pile be placed to develop an allowable load of 50 tons per pile with a factor of safety of 3?

Solution:
For a factor of safety of 3, the ultimate capacity of the pile must be $3\,(50) = 150$ tons. For cohesion $c = 0.75$ tsf, or 1.5 ksf, Equation 9.9 states $\alpha = 0.5$. The total bearing capacity is the sum of the end-bearing, Equation 9.7, and the friction capacity, Equation 9.10. $Q_b = A_b\,c\,N_c = 1.77\,(9)\,0.75 = 12$ tons and $Qs - \alpha\,c_u\,P\,L$ and $(150 - 12) = 0.5\,(0.75)\,4.71\,(L)$. Therefore, $L = 39.0$ ft.

EXAMPLE 9.7
A 40-cm diameter prestressed concrete pile has been driven 20 meters into a medium-dense sand deposit having a uniform density and an average effective shear strength, $c = 0$ and $\phi' = 32°$. As a result of a field pile load test on the single pile, it was found that the allowable load for a settlement of 16 mm was 500 kN. A nine-pile group of the same piles, spaced 160 cm center-to-center, was driven to the same depth in the same sand deposit. If the pile group is loaded to 4500 kN, what is the expected settlement of the group?

Solution:
Assuming the nine-pile group is formed in a square with three piles per side, the width B of the group is 280 cm. Then, using Equation 9.11, the expected settlement of the pile group is

$$S_g = S_i\,\sqrt{B/D} = 16\,\sqrt{360/40} = 48 \text{ mm}$$

EXAMPLE 9.8
The pressure-grouted piles of Problem 9.6 are to be formed into a nine-pile group to replace the pier of Problem 9.5. If the grouted piles are drilled only to the pier depth of 25 feet, what is the allowable bearing pressure for the group? How does this compare with the 10-foot diameter pier?

Solution:
The ultimate base-bearing capacity of a single pile, from Problem 9.6 is 12 tons. The ultimate skin friction capacity, from Equation 9.10, is $Qs = \alpha c_u P L$ and $Q_s = 0.5 (0.75) 4.71 (25) = 56.2$ tons. Therefore, the ultimate bearing capacity of a pile founded at 25 feet is $12 + 56.2 = 68.2$ tons. The group efficiency factor, from Figure 9.6, is 0.79 and the ultimate load capacity of the group is $n G_E Q_{ult} = 9 (0.79) 68.2 = 485$ tons. With a factor of safety of 3, the allowable load on the pile group is $485/3 = 162$ tons. The allowable load on the pier of Problem 9.5 is 204 tons.

EXAMPLE 9.9

In preparation for the pile load test of Problem 9.2, it was decided that the test pile depth would be determined by a pile-driving formula. The pile-driving hammer had an energy of 16,000 ft-lb. Using the Danish formula of Equation 9.12, to what penetration resistance (blows per inch) should the test pile be driven?

Solution:
Because the pile length is under the radical in Equation 9.13, it is necessary to solve Equation 9.12 for every increase in pile length as it is being driven. For this example, the solution will be derived for a pile length of 50 feet. The following factors are entered into Equation 9.12:

Pile length $L = 50$ feet $= 600$ inches,

Hammer energy $W_H H = 16,000$ ft-lbs $= 192,000$ in-lbs,

Pile area $A = 113.1$ in^2, and

Modulus of elasticity of steel $= 29,000,000$ psi

From Equation 9.13, the elastic compression per hammer drop is

$$S_e = \sqrt{\frac{2W_H HL}{AE}} = \sqrt{\frac{2 \times 192,000 \times 600}{11.31 \times 29,000,000}} = 0.265 \text{ in.}$$

Then, transposing Equation 9.12,

$$S = \frac{W_H H}{Q_{dy-ult}} - 0.5 \times S_e = \frac{192,000}{300,000} - \frac{0.265}{2} = 0.51 \text{ in./drop}$$

In practice, this then would be measured in the field as $1 / 0.51 = \sim 2$ blows (drops) per inch of permanent set of the pile.

REFERENCES

1. Peck, R. B., Hanson, W. E., and Thornburn, T. H. 1974. *Foundation Engineering.* 2nd ed., p. 215. New York: John Wiley & Sons.
2. NAVFAC DM 7.02. September 1986. "Foundations and Earth Structures," Naval Facilities Engineering Command, Department of the Navy. Alexandria, VA.

3. ASTM. 1999. "Natural Building Stones; Soil and Rock; Geotextiles," *1999 Annual Book of ASTM Standards*, vol. 04.08, American Society for Testing and Materials, Philadelphia, PA.
4. Davisson, M. T. 1973. "High Capacity Piles." *Innovations in Foundation Construction*. pp 81-112. Chicago: Soil Mechanics Division, Illinois Section, American Society of Civil Engineers.
5. Department of the Army. January 1991. "Design of Pile Foundations." *Engineer Manual EM 1110-2-2906*, U. S. Army Corps of Engineers, Washington, DC, 15 (also available as Technical and Engineering Design Guides as Adapted from the U.S. Army Corps of Engineers, No. 1, from the American Society of Civil Engineers, New York).
6. Terzaghi, K., Peck, R. B., and Mesri, G.1996. *Soil Mechanics in Engineering Practice*. 3rd ed. New York: John Wiley & Sons.
7. Nordlund, R. L. 1963. "Bearing Capacity of Piles in Cohesionless Soils." *Journal of the Soil Mechanics Division*, Vol. 89 No. SM3, pp. 1-35. American Society of Civil Engineers. New York.
8. Poulos, H. G. and Davis, E. H. 1980. *Pile Foundation Analysis and Design*.New York: John Wiley & Sons.
9. Department of the Army. October 1992. "Bearing Capacity of Soils," *Engineer Manual EM 1110-1-1905*, U. S. Army Corps of Engineers, Washington, DC.(also available as Technical and Engineering Design Guides as adapted from the U.S. Army Corps of Engineers, No. 7, from the American Society of Civil Engineers, New York).
10. Wang, S. T. and Reese L. C. 1991. "Analysis of Piles under Lateral Load—Computer Program 624P for the Microcomputer," *Report No. FHWA-SA-91-002*. Available from U. S. Department of Transportation, Federal Highway Administration, Office of Implementation, McLean, VA.
11. Sörensen, T. and Hansen, B. 1957. "Pile Driving Formulae—An Investigation Based on Dimensional Considerations and a Statistical Analysis." *Proceedings of the 4th International Conference on Soil Mechanics and Foundation Engineering*. Vol. 2, pp 61-65. London.

CHAPTER 10
RETAINING STRUCTURES

Because of its unbalanced weight, when a section of the earth is excavated and the cut surface is vertical or nearly vertical, the remaining earth tends to move laterally and downward into the excavation. This motion can be restrained by the internal shear strength of the material or by the resisting *lateral pressure* of a retaining structure. The lateral earth pressure is the significant consideration in the design of retaining walls, sheet pile walls, braced and unbraced excavations, bulkheads, buried anchorages, earth or rock pressure on tunnel walls, and grain pressure on the sides of bins and silos.

TYPES OF RETAINING STRUCTURES

Retaining structures must be designed to resist the tendency of the lateral pressure to cause overturning and sliding and must not overload the foundation soils or rock. They can be classified in several ways. One grouping subdivides walls as to their rigidity and flexibility as shown in Figure 10.1. In addition, walls are further subdivided as to their gravitational and structural qualities (are individual subdivisions a, b, c, d, e, f, g, h and i in this figure).

Gravity Retaining Walls

Gravity walls depend on their mass to resist the lateral forces. The simplest case is a concrete or rock mass, Figure 10.1a, when the base of the mass is large compared to its height. Rock can be retained in wire baskets, or cylinders called gabions, Figure 10.1b, which can be stacked at a steeper face angle than the unrestrained rock can safely assume. By using metal strips or geosynthetics as reinforcements, Figure 10.1c, sand backfill can be formed into a coherent mass that resists lateral pressures. Facing blocks are often used to restrain the outermost volume of sand that is not reinforced.

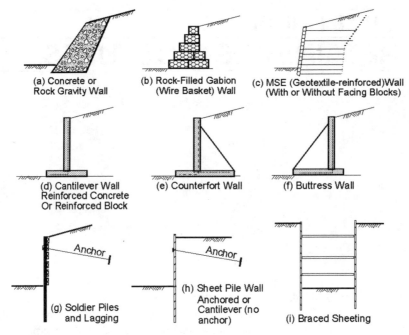

FIGURE 10.1 Types of retaining wall structures.

Structural Retaining Walls

To reduce the mass of a gravity wall, concrete walls can be formed into cantilever walls, Figure 10.1d, that rely on the steel reinforcement of the stem and base to resist lateral forces. To provide additional structural strength, cantilever walls can also be reinforced with a series of closely spaced counterforts, Figure 10.1e, on the soil side, or by a series of closely spaced buttresses, Figure 10.1f, on the outside.

Flexible Retaining Walls

Comparatively thin wall sections can be formed by driving closely spaced steel piles into the excavated area and filling the intervening spaces with wood or concrete lagging, Figure 10.1g. Another technique is to drive long, narrow, interlocking sheets of shaped steel or vinyl plastic or fiberglass as a continuous wall. Resistance to the lateral pressure is powered by the embedded sheet pile acting as a cantilever wall or as an anchored wall, Figure 10.1h. Sheet piling or wood lagging can also be used for temporary bracing, Figure 10.1i for an open cut.

LATERAL EARTH PRESSURE

A vertical plane in an earth mass with a horizontal surface is acted on by a lateral pressure that increases with depth as a direct function of the vertical pressure. This may be stated as

$$\sigma_h = K_0 \sigma_v \qquad (10.1)$$

where σ_v = vertical pressure
K_0 = coefficient of earth pressure at rest, generally about 0.4 to 0.6.
σ_h = lateral pressure

If the soil to one side of the vertical plane is removed to a depth H, the soil will tend to move into the excavation, restrained only by its internal shearing resistance (see Chapter 11). If the internal shear strength is not sufficient, a retaining structure may be placed at the location of the cut surface. The combination of internal shear strength and the restraining pressure of the structure resists the lateral pressure of the soil. This pressure on a retaining structure is a function of the relative movement between the *retaining* structure and the *restrained* soil. If the soil is permitted to move outward, toward the excavation, the lateral earth pressure on the structure will reach a minimum when the full shear strength of the soil is mobilized and the soil is in a state of plastic equilibrium, on the verge of failure. This is called *active earth pressure*. Conversely, if the retaining structure is forced inward, *toward* the soil, another state of plastic equilibrium is reached with the full shear strength of the soil mobilized. This maximum resistance is called *passive earth pressure*.

Theories have been advanced to estimate the active and passive earth pressure against retaining walls. The ones used most extensively in engineering practice were formulated by Coulomb [1] and Rankine [2]. The lateral pressure theories most applicable to the analysis and design of earth retaining structures are those that deal with *active* earth pressure.

ACTIVE EARTH PRESSURE

Coulomb, in 1776, developed one of the earliest earth pressure theories. In order to do so, he made several important assumptions:

a. The soil is homogeneous and isotropic and possesses both cohesion and internal friction.
b. The *rupture* surface is a plane surface through the heel of the wall (it is really a curved surface, but this is not normally a serious concern except in the case of earth passive earth pressure).
c. The backfill surface may slope but it is a plane surface and not irregularly shaped.
d. The failure wedge is a rigid body undergoing translation and there is no concern with the internal forces or displacements.
e. There is friction between the wall and the soil. As the failure wedge moves with respect to the back face of the wall, a friction force is formed between the soil and the wall. This friction angle usually is termed δ.

The Coulomb theory *assumes* that the pressure against the wall increases linearly with depth so that the resultant active wall pressure P_a, shown schematically in Figure 10.2, is equal to [3]:

$$P_a = \frac{\gamma H^2}{2} K_a - 2cH\sqrt{K_a} \qquad (10.2)$$

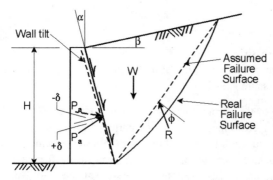

FIGURE 10.2 Assumed conditions for the Coulomb active earth pressure case.

where
$$K_a = \frac{\cos^2 \alpha + \phi}{\cos^2 \alpha \cos(\alpha - \delta)\left[1 + \sqrt{\frac{\sin(\phi + \delta)\sin(\phi - \beta)}{\cos(\alpha - \delta)\cos(\alpha + \beta)}}\right]^2} \quad (10.3)$$

Typical values of wall friction, for use in the practical design of retaining walls, are given in Table 10.1.

TABLE 10.1 Ultimate Friction Factors and Adhesion for Dissimilar Materials [4]

Interface Materials (Friction)	Friction Angle, $\delta°$
Mass concrete on the following foundation materials:	
Clean sound rock	35
Clean gravel; gravel-sand mixtures, coarse sand	29–31
Clean fine to medium sand; silty medium to coarse sand, silty or clayey gravel	24–29
Clean fine sand; silty or clayey fine to medium sand	19–24
Fine sandy silt; nonplastic silt	17–19
Very stiff and hard residual or preconsolidated clay	22–26
Medium stiff and stiff clay and silty clay	17–19
Steel sheet piles against the following soils:	
Clean gravel; gravel-sand mixture, well-graded rock fill with spalls	22
Clean sand; silty sand-gravel mixture; single size hard rock fill	17
Silty sand; gravel or sand mixed with silt or clay	14
Fine sandy silt; nonplastic silt	11
Formed concrete or concrete sheet piling against the following soils:	
Clean gravel; gravel-sand mixture; well-graded rock fill with spalls	22–26
Clean sand; silty sand-gravel mixture; single size hard rock fill	17–22

TABLE 10.1 Ultimate Friction Factors and Adhesion for Dissimilar Materials [4] *(continued...)*

Interface Materials (Friction)	Friction Angle, $\delta°$
Formed concrete or concrete sheet piling against the following soils (continued...):	
Silty sand; gravel or sand mixed with silt or clay	17
Fine sandy silt; nonplastic silt	14
Various structural materials:	
Masonry on masonry; igneous and metamorphic rocks	
Dressed soft rock on dressed soft rock	35
Dressed hard rock on dressed soft rock	33
Dressed hard rock on dressed hard rock	29
Masonry on wood (cross grain)	26
Steel on steel at sheet pile interlocks	17

Interface Materials (Cohesion)	Adhesion C_A, psf
Very wet cohesive soils (0–250 psf)	0–250
Soft cohesive soil (250–500 psf)	250–500
Medium stiff cohesive soil (500–1000 psf)	500–750
Stiff cohesive soil (1000–2000 psf)	750–950
Very stiff cohesive soil (2000–4000 psf)	950–1300

RANKINE'S THEORY

In 1857, Rankine simplified the Coulomb theory by disregarding wall friction. Setting $\delta = 0°$ in Equation 10.3, the general Rankine theory states that, for use in Equation 10.2:

$$K_a = \frac{\cos^2(\alpha + \phi)}{\cos^3\alpha \left[1 + \sqrt{\frac{\sin\phi \sin(\phi - \beta)}{\cos\alpha \cos(\alpha + \beta)}}\right]^2} \tag{10.4}$$

For the simple case where the wall is vertical ($\alpha = 90°$) and the backfill is horizontal ($\beta = 0°$):

$$K_a = \frac{1 - \sin\phi}{1 + \sin\phi} = \tan^2\left(45 - \frac{\phi}{2}\right) \tag{10.5}$$

In practical design situations, the calculation of the active pressure on a retaining wall can be made using either the Coulomb theory or the Rankine theory. In both cases, the point of application of the force is at the centroid of the pressure triangle, at *H/3* from the bottom. The triangular pressure distribution is similar to the pressure that results from loading with a fluid. For this reason, the magnitude of the active pressure on the back of a retaining wall often is referred to as *equivalent fluid pressure*.

Figure 10.3 indicates the differences in active lateral pressure calculated by two methods:

a. The *Coulomb* active pressure is applied directly to the back of the wall at an angle from the horizontal of $\delta + \alpha$.

b. The *Rankine* active pressure is applied to a vertical plane extending upward from the heel of the wall, inclined at β, the angle of the backfill, and extending over a height H_1. The weight of the soil W_s between the wall and the vertical plane adds to the weight of the concrete wall, W_c.

COHESIVE SOILS

If a shallow cut is made in a cohesive soil, the bank may remain vertical without any support. The *critical height* H_c, the height to which an unsupported vertical cut may be made in cohesive soil, is derived from Equation 10.2 by finding the height at which the *total* active pressure, the sum of the positive and negative earth pressures, equals zero. The critical height for cohesive soil with $\phi = 0°$ is:

$$H_o = \frac{4c}{\gamma} = 2z_o \qquad (10.6)$$

where z_0 is the depth at which the pressure against a retaining wall is zero, where the active pressure diagram starts.

SURCHARGE LOADING

Loads on the surface of the backfill add to the lateral earth pressure in addition to active force. With a *uniform* surcharge load, there will be an additional *uniform* pressure of $q\,K_a$ applied at $H/2$ in the same direction as the active pressure. Horizontal pressures due to point, line, and strip loads were discussed in Chapter 4 and shown in Figures 4.8, 4.9, and 4.10.

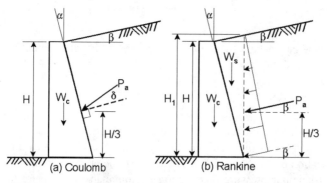

FIGURE 10.3 Active forces on a gravity retaining wall with granular backfill according to (a) the Coulomb theory and (b) the Rankine theory.

ACTIVE LATERAL PRESSURE DUE TO EARTHQUAKE FORCES

A design procedure for employing the Mononobe-Okabe calculations of pseudostatic earthquake accelerations to the Coulomb lateral earth pressures was presented by Seed and Whitman [5] in terms of a cohesionless backfill that was not subject to liquefaction. In the procedure, the total active thrust due to the earthquake forces is in a form similar to that for static conditions:

$$P_{ae} = \frac{\gamma H^2}{2}(1 - k_v)K_{ae} \tag{10.7}$$

where

$$K_{ae} = \frac{\cos^2(\phi - \alpha + \psi)}{\cos\psi \cos^2\alpha \cos\left(\delta + \alpha + \psi\right)\left[1 + \sqrt{\frac{\sin(\delta + \phi)\sin(\phi - \beta - \psi)}{\cos(\delta + \alpha + \psi)\cos(\beta - \alpha)}}\right]^2} \tag{10.8}$$

in which $\psi = \tan^{-1}\frac{k_h}{1 - k_v}$ = seismic inertia angle (10.9)

where k_h = Horizontal ground acceleration in g's
k_v = Vertical ground acceleration in g's.

Seed and Whitman recommended that the dynamic component be applied at $0.6H$ and the static component continue to be applied at $H/3$, as shown in Figure 10.4, by separating the two components as:

$$\Delta P_{ae} = P_{ae} - P_a \tag{10.10}$$

They further recommended that the horizontal ground acceleration used in the seismic inertia angle, Equation 10.9, be 85 percent of the peak horizontal acceleration for the design earthquake and that the vertical acceleration be taken as two-thirds of the horizontal.

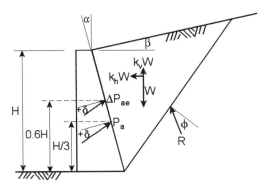

FIGURE 10.4 Forces acting on the active wedge using the Seed and Whitman earthquake pseudostatic force method.

PASSIVE EARTH PRESSURE

The Coulomb passive earth pressure P_p shown schematically in Figure 10.5 is equal to

$$P_p = \frac{\gamma H^2}{2} K_p + 2c\sqrt{K_p} \tag{10.11}$$

where

$$K_p = \frac{\cos^2(\alpha + \phi)}{\cos^2\alpha \cos(\delta - \alpha)\left[1 + \sqrt{\dfrac{\sin(\phi - \delta)\sin(\phi + \beta)}{\cos(\delta - \alpha)\cos(\beta - \alpha)}}\right]^2} \tag{10.12}$$

The equivalent Rankine passive pressure for a wall with no friction can be derived by solving Equations 10.11 and 10.12 using $\delta = 0°$. For the simple case where the wall is vertical ($\alpha = 90°$) and the backfill is horizontal ($\beta = 0°$), the Rankine passive pressure coefficient is:

$$K_p = \frac{1 + \sin\phi}{1 - \sin\phi} = \tan^2\left(45 + \frac{\phi}{2}\right) \tag{10.13}$$

As can be seen from Figure 10.5, the Coulomb-Rankine methods that assume a planar failure surface through the heel of the wall underestimate the passive pressure by a significant amount. If the actual passive earth pressure is needed for a design configuration, it is necessary to resort to other methods such as the logarithmic spiral or the trial wedge to achieve greater accuracy. These analysis techniques can be found in standard geotechnical engineering textbooks.

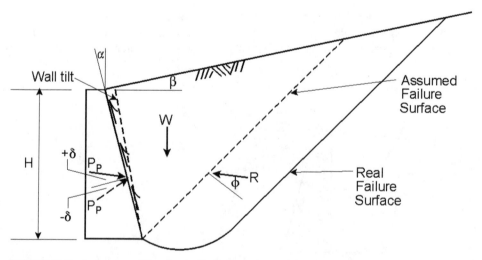

FIGURE 10.5 Assumed conditions for the Coulomb passive earth pressure case.

DESIGN OF RIGID RETAINING WALLS

A retaining wall must be stable against overturning, sliding, or foundation failure. The forces acting on a rigid retaining wall are typified by the forces acting on a cantilever retaining wall as shown in Figure 10.6.

The design of a rigid retaining wall generally comprises the following steps [6]:

1. Assemble all pertinent information including subsoil conditions, backfill sources and types, required height of wall, and surcharge.
2. Select the type and tentative proportions of the wall.
3. Calculate the static earth pressure, surcharge pressure, and earthquake loading.
4. Analyze overturning and maximum soil pressure.
5. Analyze sliding and the need for a foundation key.
6. Analyze global stability.
7. Design structural elements and backfill drainage.

Safety factors for walls should be designed to prevent overturning of 1.5 for granular backfill and of 2.0 for cohesive backfill. The factor of safety against sliding should be at least 1.5. The calculation of the pressures on retaining walls 20 feet high or less can be facilitated by design charts such as that shown in Figure 10.7. The following typical proportions may be used to select the initial trial section [6]:

1. Gravity walls are primarily trapezoidal in section; the base generally ranges between 30 to 40 percent of the height.
2. Cantilever walls generally have (a) a footing width of 0.4 to 0.67 times the height, (b) a stem width of 8 in. minimum with 12 inches normal, and (c) a toe projection of 0.1 to 0.12 times the footing width.

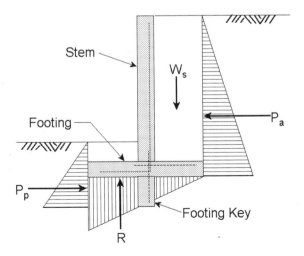

FIGURE 10.6 Forces acting on a cantilever retaining wall.

3. For counterfort walls of moderate height the counterforts may be spaced as far as two-thirds of the height, while for walls higher than 30 feet, the spacing may be reduced to one-half the height, but not less than about 8 feet.

DESIGN OF MSE RETAINING WALLS

A mechanically stabilized earth (MSE) retaining wall is essentially a gravity wall. The combination of stacked blocks and horizontal reinforcement using metal strips or geosynthetic sheets provides sufficient coherence to the reinforced backfill. This coherence insures that it reacts to external active pressure forces as an intact block. The interior of the block is acted on by internal active pressure forces that cause tension in the reinforcement and apply pressure to the individual blocks. The main modes of failure, for which analyses are made in the design procedure, are shown in Figure 10.8.

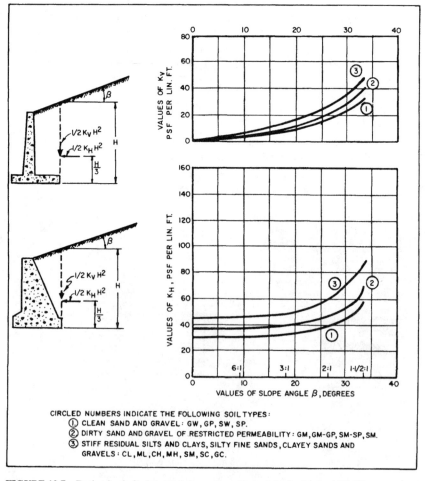

FIGURE 10.7 Design loads for low retaining walls with straight sloping backfill [4].

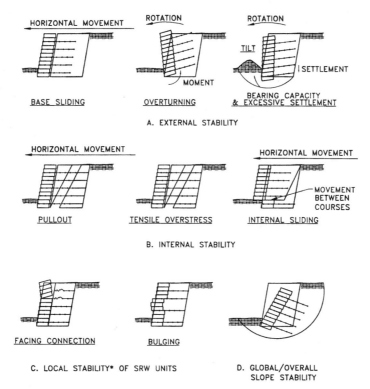

FIGURE 10.8 Main modes of failure for mechanically stabilized earth walls. (*Courtesy of the National Concrete Masonry Association* [7])

The recommended methodology for design of MSE walls is given in a National Concrete Masonry Association publication [7]. The essentials of the design procedure are:

1. Assemble all pertinent information: subsoil conditions, backfill sources and types, the required height of the wall, and the surcharge.
2. Select the type and tentative proportions of the wall and reinforcement.
3. Investigate *external* stability (Figure 10.8a) due to static lateral earth pressure, surcharge pressure, and earthquake loading using the Coulomb equations for rigid walls.
4. Investigate *internal* stability (Figure 10.8b) due to static lateral earth pressure, surcharge pressure, and earthquake loading using the Coulomb equations. Calculate the forces tending to cause pullout and tensile overstress of the individual reinforcement strips or sheets and tending to cause sliding between reinforcement courses.
5. Investigate the local stability of facing connections and bulging of individual facing blocks (Figure 10.8c).
6. Investigate global slope stability (Figure 10.8d) using the methodology to which you will be introduced in Chapter 11.

DESIGN OF FLEXIBLE RETAINING WALLS

There are two basic types of sheet pile walls: cantilevered and anchored. Cantilever walls rely on the firm embedment of the pile in the ground, and on the pile stiffness to resist the applied loads as a true cantilever beam. Cantilever walls using steel sheet piling are restricted to a maximum height of about 15 feet (4.6 m). For higher walls, or to use smaller sections, the wall may be provided with an earth anchor near the top of the piling. Anchored steel sheet piling may be used for heights up to about 35 feet (10.5 m).

The design of sheet pile walls requires several successive analyses [8]:

1. Evaluation of the forces and lateral pressures that act on the wall.
2. Determination of the required depth of piling penetration.
3. Calculation of the maximum bending moments in the piling.
4. Calculation of the stresses in the wall and selection of the appropriate piling section.
5. For anchored piles, design of a waling and anchorage system.

Lateral Pressures. The gross and net active and passive pressures on a sheet pile wall are shown, for a homogeneous soil, in the resultant earth pressure diagram of Figure 10.9. In that diagram, although not shown, the unit weight of soil to be used in calculating earth pressure is either the wet unit weight γ or the submerged unit weight γ_{sub}.

In Figure 10.9, it is assumed that the piling is rigid and rotates about point O. The total active earth pressure acting against the entire pile is indicated by ABC. The total passive earth pressure available against the lower section of the pile is EFG. The active pressure at the dredge line ALE reduces the passive pressure by a constant amount EL, resulting in a net active pressure P_1 of ALO_1 and a net passive pressure of O_1GC. The total available passive pressure for resisting the movement of the bottom of the pile toward the high side is AKC, which is reduced by the available active pressure from the low side to a net value of AJC. The net passive pressures from both sides of the pile reduce to P_2 and P_3. Static equilibrium requires the sum of net forces from both sides to equal zero and the sum of moments about any point also must equal zero. The use of an anchorage tied to the pile near its top simply introduces another force into the system.

Depth of Piling Penetration. In the analysis of sheet piling driven into granular soils, it is customary to assume a trial depth of penetration from the first calculation. Teng [6] suggests a depth of penetration in terms of the height H above the dredge line:

Soil	Depth of Penetration
Very loose sand	$2.0 H$
Loose sand	$1.5 H$
Medium dense sand	$1.25 H$
Dense sand	$1.0 H$
Very dense sand	$0.75 H$

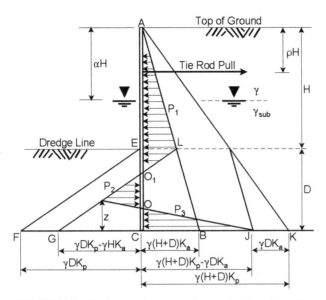

FIGURE 10.9 Resultant earth pressure diagram for sheet pile wall.

Teng also suggests adding 20 to 40 percent to the calculated depth of penetration to provide a factor of safety of about 1.5 to 2.0. The maximum bending moment, which occurs at the point of zero shear, should be calculated without the added penetration depth.

Maximum Bending Moments. The calculation of pressures and bending moments for a sheet pile wall is a tedious process. The digital computer is a useful calculation aid that permits the designer to easily evaluate "what-if" scenarios for different backfill, water level, wall height, and other wall configurations.

As a design aid, the U. S. Steel Design Manual [8] contains charts, included as Figures 10.10 through 10.13, to assist in the design of normally encountered situations. The charts permit determination of the depth ratio D/H and of the maximum bending moment as a function of the active pressure. The charts can be used alone or checked by calculators or computers. Examples of the use of the charts are contained in the Example Problems below.

Anchorage Capacity. Anchorages, also known as deadmen, anchor blocks, or anchor beams, may be constructed near the surface or at great depth and may be short or long. Sketches to accompany this discussion are shown in Figure 10.14.

1. If the depth of the top of the continuous (very long compared to its height) deadman h is less than about one-third to one-half the depth to the bottom of the deadman H, Figure 10.14 (a), the capacity is nearly the same as if the deadman extends to the surface. Then $T_{ult} = P_p - P_a$, where T_{ult} is the ultimate capacity of the deadman, and P_p and P_a are the passive and active earth pressures.

2. For a short deadman near the surface in granular soil, Teng [6] recommends:

$$T_{ult} \leq L(P_p - P_a) + 1/3 K_a \gamma \left(\sqrt{K_p} + \sqrt{K_a}\right) H^3 \tan \phi \qquad (10.14)$$

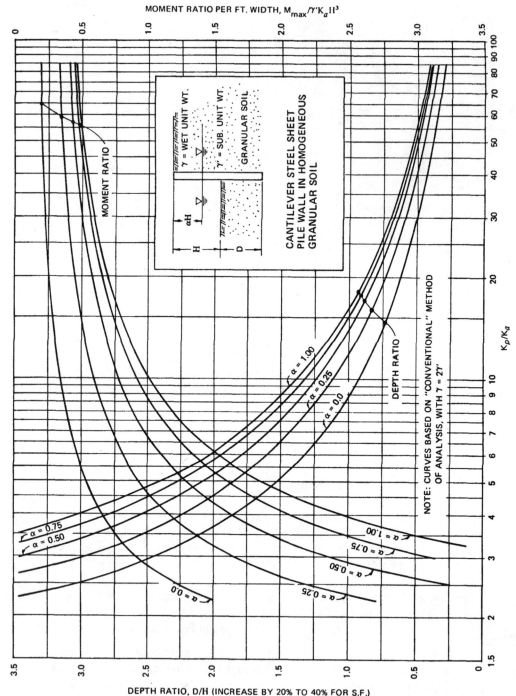

FIGURE 10.10 Design chart for cantilever sheet pile wall in homogeneous granular soil. (*Courtesy of USX Corporation, formerly U.S. Steel Corporation* [8])

10.14

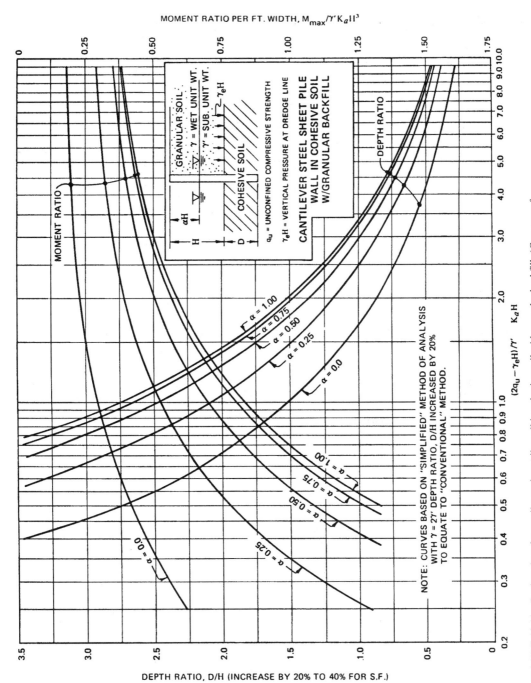

FIGURE 10.11 Design chart for cantilever sheet pile wall in cohesive soil with granular backfill. (*Courtesy of USX Corporation, formerly U.S. Steel Corporation* [8])

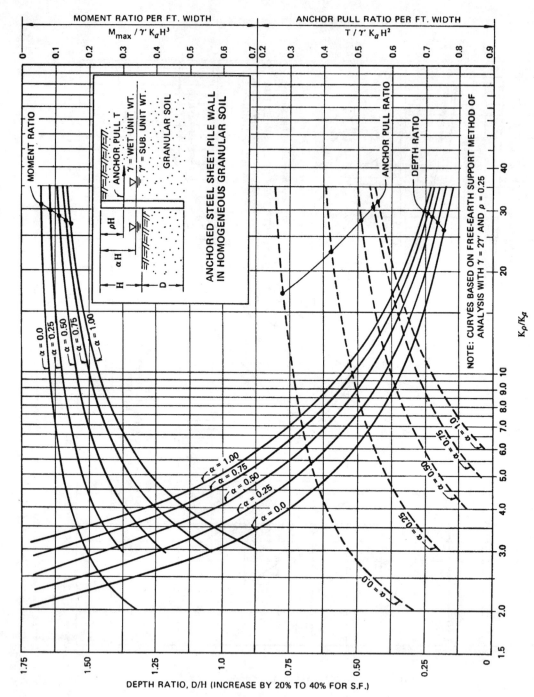

FIGURE 10.12 Design chart for anchored sheet pile wall in homogeneous granular soil. (*Courtesy of USX Corporation, formerly U.S. Steel Corporation* [8]).

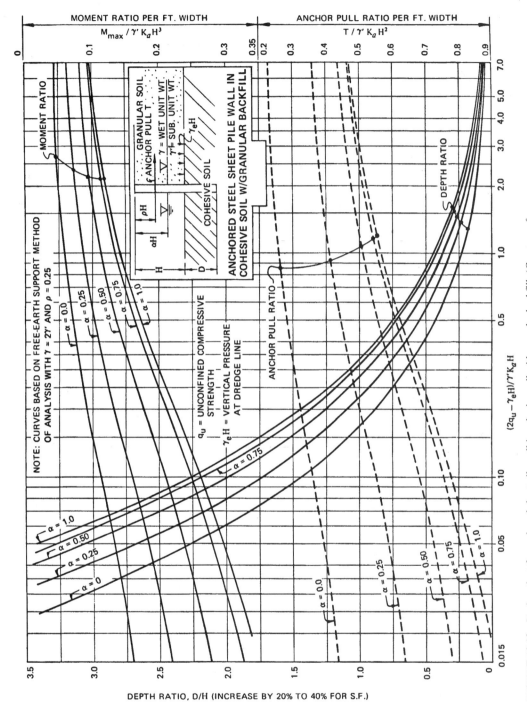

FIGURE 10.13 Design chart for anchored sheet pile wall in cohesive soil with granular backfill. (*Courtesy of USX Corporation, formerly U.S. Steel Corporation* [8])

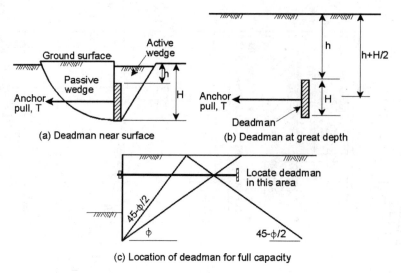

FIGURE 10.14 Capacities of deadmen (a) near the surface and (b) at great depth, and (c) location of the deadman for full capacity (adapted from Teng [6])

where L = Length of deadman
 P_p, P_a = Total passive and active earth pressure
 K_o = Coefficient of earth pressure at rest, may be taken as 0.4
 γ = Unit weight of soil
 K_p, K_a = Coefficients of passive and active earth pressure.

3. The ultimate capacity of a deadman at great depth ($h > H$), Figure 10.14(b), is about equal to the bearing capacity of a footing the size of the deadman and located at the mid-depth of the deadman.

Figure 10.14(c) shows the optimum location for a deadman to achieve full capacity. If the deadman is located within the active pressure wedge there is no resistance available.

DESIGN OF BRACED EXCAVATIONS

Temporary or permanent cuts or trenches are often cut into soils to permit placement of such facilities as sewer lines and other types of pipe lines. If a shallow cut is made into clay, theoretically the cut will stay open without bracing to the critical depth defined in Equation 10.6. We know that shallow cuts will not remain vertical in cohesionless materials. Therefore, cross-bracing as shown in Figure 10.1i is used. Normally the bracing is placed in sequence, moving downward as the excavation is made. As a result, the deflection of the braced wall is not the same as the deflection for a retaining wall. It is bowed because the top is restrained. Extensive field measurements of actual bracing pressures have led to the formulation of empirically sound pressure diagrams. Pressure distributions recommended by Terzaghi, Peck, and Mesri [9] for use in the design of strut loads are shown in Figure 10.15.

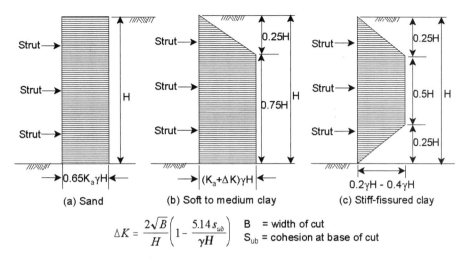

FIGURE 10.15 Suggested apparent earth pressure diagrams for design of struts in open cuts in (a) sand, (b) soft to medium clays, and (c) stiff-fissured clays.

$$\Delta K = \frac{2\sqrt{B}}{H}\left(1 - \frac{5.14 s_{ub}}{\gamma H}\right)$$

B = width of cut
S_{ub} = cohesion at base of cut

EXAMPLE PROBLEMS

EXAMPLE 10.1

A 15-ft high gravity retaining wall has a granular backfill with a unit weight of 125 lb/ft³. The water table is below the bottom of the wall. Determine the Coulomb and Rankine active pressures, the per foot of length of the wall, and compare the results for walls having the following characteristics (for δ use ⅔ φ):

	Backfill ϕ	Wall angle α	Backfill angle β
a.	36	15	20
b.	36	15	0
c.	36	0	20
d.	36	0	0

Solutions:
Use Equation 10.2 with cohesion $c = 0$. Solve for Coulomb active pressure using Equation 10.3 and for Rankine active pressure using Equation 10.4. Determine the direction of the force application from Figure 10.3.

	Backfill ϕ	Wall angle α	Backfill angle β	Pa, Coulomb lbs/ft wall	Pa, Rankine lbs/ft wall	Ratio: Rankine to Coulomb
a.	36	15	20	6990	6840	0.98
b.	36	15	0	5080	5260	1.04
c.	36	0	20	4320	4600	1.06
d.	36	0	0	3305	3650	1.10

EXAMPLE 10.2

A 15-ft high gravity retaining wall has a compacted silty clay backfill with a unit weight of 120 lb/ft³. Its cohesion $c = 300$ psf, and $\phi = 0°$. The water table is below the bottom of the wall. What is the total Coulomb earth pressure against the wall?

Solution:
Equation 10.6 indicates that for the clayey soil the depth at which the active pressure is zero is $2c/\gamma = 2(300)/120 = 5.0$ ft. Therefore the active pressure force triangle starts at a depth of 5 ft.

$$P_a = 0.5\, K_a\, (H - z_0)^2 = 0.5 \times 1 \times 120 \times 10^2 = 6000 \text{ lb/ft of wall}$$

EXAMPLE 10.3

The gravity retaining wall of Example Problem 10.1 has a uniform surcharge of 500 lbs/ft² applied to the top of the backfill from the wall outward for a distance of at least 15 feet. What is the Coulomb lateral pressure on the wall?

Solution:
The Coulomb active earth pressure coefficient for $\phi = 36°$ is $K_a = 0.260$. The uniform surcharge applies an additional *uniform* pressure of qK_u which acts at $H/2$. The active earth pressure is $P_a = 3305$ lbs/ft plus $P_q = 500\,(0.260) = 130$ lbs/ft for a total active earth pressure of $P_a + P_q = 3435$ lbs/ft of wall.

EXAMPLE 10.4

The gravity retaining wall of Example Problem 10.1d is subjected to an earthquake force of 0.4 g. What is the active earth pressure for this seismic loading?

Solution:
Using only 85 percent of the earthquake force of 0.4 g, the horizontal acceleration component $k_h = 0.85\,(0.4) = 0.34$ g and the vertical component k_v as ⅔ k_h in Equation 10.9, the seismic inertia angle $\psi = 27°$. Solving Equation 10.8 for the active pressure coefficient due to earthquake loading, $K_{ae} = 0.704$ and the total active earth pressure is, from Equation 10.7,

$$P_{ae} = 0.5\,(125)\,15^2\,(1 - 0.23)\,0.704 = 7623 \text{ lbs/ft of wall}$$

EXAMPLE 10.5

A concrete cantilever retaining wall is 12 feet high with a 7 ft wide base. The backfill is a silty fine sand with a unit weight of 120 lb/ft³. The backfill slopes upward from the rear of the stem of the wall at a slope of 3H : 1V (18.4°). Estimate the active earth pressure acting on the wall.

Solution:
Assume the heel width of the wall is 4 ft. Then the vertical height from the heel to the backfill will be $H_1 = 12 + 4\,(1/3) = 13.3$ ft. From Figure 10.7, the vertical component k_v of active pressure is about 10 psf and the horizontal component k_h is about 50 psf. The inclined active pressure coefficient is then 51 psf. The active earth pressure is then:

$$P_a = 0.5\ (51)\ (13.3)^2 = 4510\ \text{lbs/ft of wall}$$

EXAMPLE 10.6

An anchored steel sheet pile wall is 16 feet high above the dredge line. The anchor to a deadman is applied at a depth of 4 feet below the top of the pile ($\rho = 0.25$). The soil behind the wall and the foundation soil is a medium sand having a wet unit weight of 125 lbs/ft^3 and a friction angle of $\phi = 35°$. The coefficient of friction between the wall and the sand is $\delta = 0.5\ \phi = 17.5°$. The low-water level is at the dredge line. Estimate (a) the needed depth of sheet pile (at FS = 1), (b) the maximum moment on the sheet piling, and (c) the anchor pull.

Solution:
For $\phi = 35°$ and $\delta = 17.5°$, the Coulomb active earth pressure coefficient $K_a = 0.260$ and the passive earth pressure coefficient $K_p = 7.36$. The ratio $K_p / K_a = 29.9$. For Figure 10.12, $\alpha = 1.0$ and $\gamma' = 125 - 62.4 = 62.6$ pcf. Then:

a. From Figure 10.12 the depth ratio $D/H = 0.25$. The required depth is then 4 feet. However, to achieve a factor of safety of 1.5 to 2.0, add about 30 percent to the depth so that the actual depth is slightly greater than 5 feet.

b. The moment ratio $M_{max} / \gamma'\ K_a\ H^3 = 0.15$. Therefore the maximum moment $M_{max} = 62.6\ (0.260)\ 16^3\ (0.15) = 10,000$ ft-lbs.

c. The anchor pull ratio $T / \gamma'\ K_a\ H^2 = 0.57$. Therefore the maximum moment anchor pull $T = 62.6\ (0.260)\ 16^2\ (0.57) = 2375$ lbs per foot of wall.

REFERENCES

1. Coulomb, C. A. 1776. "Essai sur une Application des Règles des Maximis et Minimis à quelques Problèmes de Statique Relatifs à l'Architecture" (An attempt to apply the rules of maxima and minima to several problems of stability related to architecture), *Mem. Acad. Roy. des Sciences*, vol. 7, pp. 343-382. Paris.
2. Rankine, W. J. M. 1857. "On the Stability of Loose Earth," *Philosophical Transactions of the Royal Society*, vol. 147, part 1, pp. 9-27. London
3. Terzaghi, K. 1943. *Theoretical Soil Mechanics*. New York: John Wiley & Sons.
4. NAVFAC DM 7.02. 1986. *Foundations and Earth Structures, Design Manual DM 7.02*. Naval Facilities Engineering Command, Department of the Navy. Alexandria, VA.
5. Seed, H. B. and Whitman, R.V. 1970. "Design of Earth Retaining Structures for Dynamic Loads," *Proceedings, ASCE Specialty Conference on Lateral Stresses in the Ground and Design of Earth Retaining Structures*. American Society of Civil Engineers. New York.
6. Teng, W. C. 1962. *Foundation Design*. Englewood Cliffs, NJ: Prentice-Hall.
7. National Concrete Masonry Association. 1997. *Design Manual for Segmental Retaining Walls*, J. G. Collin, Editor, Second Edition. National Concrete Masonry Association. Herndon, VA.
8. U. S. Steel. 1975. *Steel Sheet Piling Design Manual*. United States Steel Corporation. Pittsburgh, PA.
9. Terzaghi, K., Peck, R. B., and Mesri, G. (1996). *Soil Mechanics in Engineering Practice*, Third Edition. New York: John Wiley & Sons.

CHAPTER 11
SLOPE STABILITY

The soil or rock in a slope exist in a state of equilibrium between gravity forces tending to move the material down the slope and the internal shearing resistance of the material. A slope failure occurs when the force tending to cause rupture exceeds the resisting force. The overstressing of a slope or reduction in shear strength may cause displacements that may be very slow or very rapid and progressive.

Extremely slow movements in soils are called *soil creep*. Rapid movements of intact or nearly intact soil or rock masses are called *slides*. Rock or soil that detaches from a nearly vertical slope and descends mainly through the air by falling, bouncing, or rolling is called a *fall*. Very soft cohesive soils can fail by lateral spreading or by *mud flows*.

This chapter is concerned only with *slides* and how to anlayze their stability. In fact, it is only concerned with rotational and translational slides that can be analyzed with relatively simple and straightforward mathematical models.

TYPES OF SLOPE FAILURES

The principal causes of slope failures of slopes in soil or in rock are (a) displacement of a wedge-shaped mass along a plane of weakness, (b) translation of a wedge-shaped mass, whose length is large compared to its depth, along a plane of weakness, and (c) rotation along a curved surface that can be approximated by a circular arc. For the types of failures in natural or completed soil slopes see Figure 11.1.

1. *Slope in coarse-grained soils with some cohesion.* With low ground water, failure occurs on shallow, straight or slightly curved surface. Presence of a tension crack at the top of the slope influences the failure location. With high ground water, failure occurs as a relatively shallow toe circle.
2. *Slope in coarse-grained cohesionless soil.* Stability depends primarily on ground water conditions. With low ground water, failure is by sloughing until the slope

angle flattens to the angle of internal friction. With high ground water, because of seepage pressure, stable slope is about one-half of the friction angle.

3. *Slope in soft to stiff homogeneous clay.* Failure occurs on a deep circular arc whose position is governed by slope angle and relative shear strength.
4. *Slope in stratified soil profile.* Location of the failure plane is along the planar surface and is controlled by relative strength and orientation of the strata.
5. *Depth creep movements in an old slide mass.* The strength of an old slide mass decreases with the magnitude of movement that has occurred previously due to soil remolding.

SLIDING WEDGE ANALYSIS

In stratified soil slopes of the type illustrated by Figure 11.1(4), where there is (a) a layer of low-strength material below a much stronger layer or (b) where there is a firm base below one or more lower strength layers, the lateral earth pressure concepts discussed in Chapter 10 can be used to analyze the stability of the system.

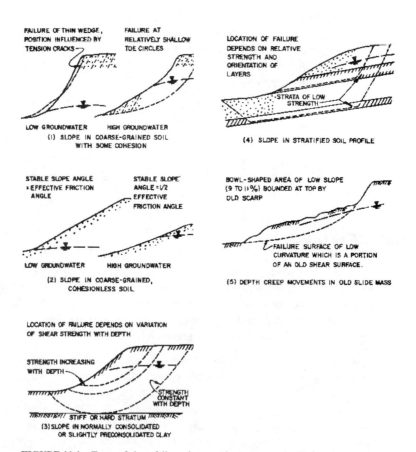

FIGURE 11.1 Types of slope failures in natural or compacted soil slopes.

Figure 11.2 shows the forces acting on a translation, or sliding wedge, type of failure. The driving force(s) on the soil mass are the active pressure wedges and the resisting force(s) are the passive pressure wedges. Depending on stratigraphy and soil strength, the central wedge may act to maintain or upset equilibrium.

Where the terms in Figure 11.2 are identified [1, p. 7.01-324] as:

P_α = Resultant horizontal force for an active or central wedge along potential sliding surfaces $a\ b\ c\ d\ e$.

P_β = Resultant horizontal force for a passive wedge along potential sliding surfaces $e\ f\ g$.

W = Total weight of soil and water in the wedge above the potential sliding surface.

R = Result of normal and tangential forces on the potential sliding surface considering the friction angle of the material.

P_w = Resultant force due to the pore water pressure on the potential sliding surface.

ϕ = Friction angle of the layer along the potential sliding surface.

c = Layer cohesion along the potential sliding surface.

L = Length of the potential the sliding surface across the wedge.

h_w = Depth below phreatic surface at the boundary of the wedge

γ_w = Unit weight of water.

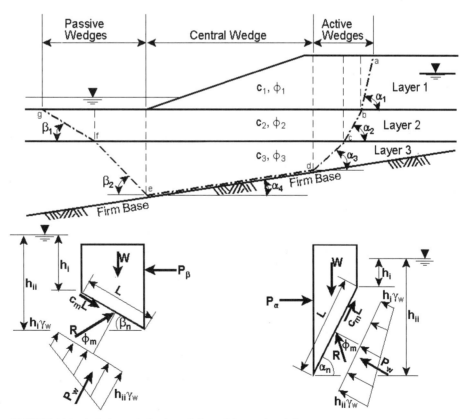

FIGURE 11.2 Stability analysis of translation (sliding wedge) failure [1].

The safety factor is defined in terms of the amount of each strength term that is mobilized by the driving forces:

$$\phi_m = \tan^{-1}\left(\frac{\tan\phi}{FS}\right) \text{ and } c_m = \frac{c}{FS} \quad (11.1)$$

Using an appropriate factor of safety FS applied as in Equation 11.1, then

$$P_\alpha = (W - c_m L \sin\alpha - P_w \cos\alpha)\tan(\alpha - \phi_m) - (c_m L \cos\alpha - P_w \sin\alpha) \quad (11.2)$$

$$P_\beta = (W + c_m L \sin\beta - P_w \cos\beta)\tan(\beta + \phi_m) + (c_m L \cos\beta + P_w \sin\beta) \quad (11.3)$$

in which

$$\tan(\alpha - \phi_m) = \frac{\tan\alpha - \dfrac{\tan\phi}{FS}}{1 + \tan\alpha \dfrac{\tan\phi}{FS}} \quad (11.4)$$

$$\tan(\beta + \phi_m) = \frac{\tan\beta + \dfrac{\tan\phi}{FS}}{1 - \tan\beta \dfrac{\tan\phi}{FS}} \quad (11.5)$$

The resultant force P_w due to pore water pressure on the potential sliding surface is calculated as:

$$P_w = \left(\frac{hw_i + hw_{ii}}{2}\right) L \gamma_w \quad (11.6)$$

The recommended [1, p. 7.01-324] procedure for stability analysis of a sliding wedge is:

1. Use $\alpha = 45° + \phi/2$ for active wedges and $\beta = 45° - \phi/2$ for passive wedges for estimating failure except for the central wedge where α is dictated by stratigraphy.
2. For a trial location and length L of the central wedge and/or for a trial factor of safety, solve for the active force P_α, using Equations 11.2 and 11.4, and for the passive force P_β, using Equations 11.3 and 11.5, for each layer for the selected safety factor.
3. The direction of the shearing resistance of the central wedge will depend on the base angle and the force should be added to either the passive forces or the active forces as appropriate. For a positive angle α_4 (as shown in Figure 11.2) the shearing resistance of the central wedge will be added to the passive forces to resist the active forces.
4. For static equilibrium, the summation of the driving and resisting forces must be equal. Because the factor of safety term appears in both the active force and passive force equations, it is necessary to repeat both sets of equations using the same FS until equivalence is reached.
5. Note that for $\phi = 0°$, the above equations reduce to

$$P_a = W \tan\alpha - \frac{c_m L}{\cos\alpha} \text{ and } P_\beta = W \tan\beta + \frac{c_m L}{\cos\beta} \quad (11.7)$$

6. The safety factor for several potential sliding surfaces may have to be computed in order to find the minimum *FS* for the given stratigraphy.

INFINITE SLOPE ANALYSIS

The sliding wedge analysis can be simplified when a layer of firm soil or rock lies parallel to a thin layer of softer material and the potential slip surfaces are very long compared to their depth. This occurs when a rock surface is parallel to the slope and there is a thin layer of soil overlying the rock. This case can be approximated by an infinite slope analysis. In this analysis, the driving forces of the uphill wedges and the resisting forces of the downhill wedges are ignored, and only the remaining *central wedge* is considered.

A slope of infinite length (no end effects), such as illustrated by Figure 11.1 (1) and (2), is shown schematically in Figure 11.3. The slope is at an angle β from the horizontal and has a thickness t normal to the slope. A water level exists at a constant depth mt (where m is a decimal fraction of t). The sliding mass has shear strength consisting of c and ϕ.

For the general case shown in Figure 11.3, the factor of safety against sliding is [2, p. 354]:

$$FS = \frac{c' + h \cos^2 \beta \, [(1 - m)\gamma_m + m\gamma'] \tan\phi'}{h \sin\beta \cos\beta \, [(1 - m)\gamma_m + m\gamma_{sat}]} \tag{11.8}$$

in which c' and ϕ' are the effective stress cohesion and friction angle.

For the simple case of a *dry sand*, where $c = 0$, $N = W \cos \beta$, and $T = W \sin \beta$, Equation 11.8 reduces to

$$FS = \frac{N \tan \phi}{W \sin \beta} = \frac{\tan \phi}{\tan \beta} \tag{11.9}$$

For this case, the factor of safety is independent of the slope height h and depends only on the angle of internal friction ϕ for the sand and the angle of the slope β. For a factor of safety $FS = 1$, the slope angle is limited to the angle of internal friction of the sand.

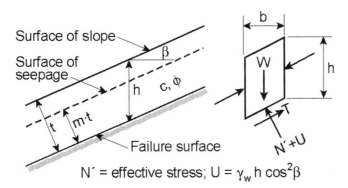

FIGURE 11.3 Stability of infinite slope with seepage. [2]

SLOPE STABILITY CHARTS FOR HOMOGENEOUS SLOPES

Theoretical solutions have been obtained [3] for the case of simple slopes without seepage entirely in homogeneous deposits. The accuracy of the charts is often as good as the accuracy with which shear strengths can be evaluated. To apply them to real conditions it is necessary to approximate the real slope with an equivalent simple and homogeneous slope [4]. A chart is shown in Figure 11.4 for determining a dimensionless stability number for a cohesive soil slope.

The assumed conditions for development of the charts were: (1) no open water outside of the slope, (2) no surcharge or tension cracks, (3) the soil is homogeneous to depth D, (4) the shear strength is derived from cohesion only and is constant with depth, and (5) failure takes place as rotation on a circular arc. The factor of safety FS for a slope of cohesion c, height H, slope angle β, unit weight γ, and with stability number N_0 is:

$$FS = N_0 \frac{c}{\gamma H} \qquad (11.10)$$

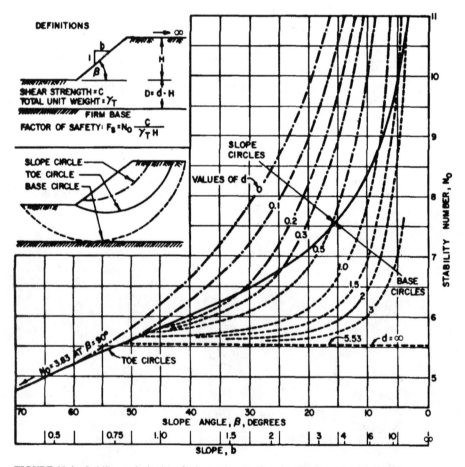

FIGURE 11.4 Stability analysis chart for homogeneous slope in cohesive soils, $\phi = 0°$ [1].

To average the shear strengths for chart analysis it is necessary to know approximately where the critical circle is located. The chart of Figure 11.5 is used to determine the location of the critical circle, after which a scale drawing should be made of the slope, the critical circle drawn to scale, and the portion of the circle that lies within each layer or zone established. Weighted averages of the cohesion for each layer can then be assigned to establish the average cohesion for the entire slope for use in Equation 11.10.

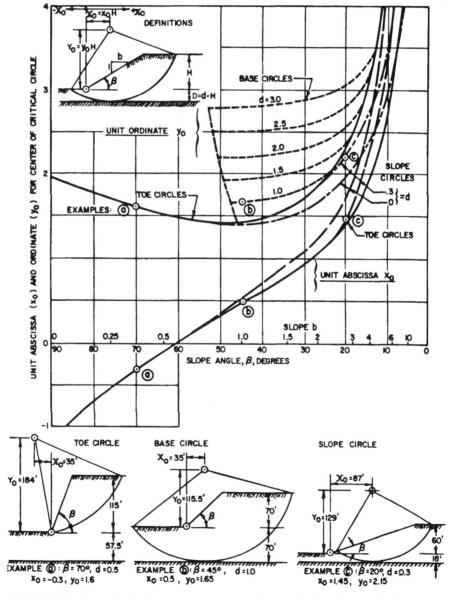

FIGURE 11.5 Center of critical circle for homogeneous slope in cohesive soils, $\phi = 0°$ [1].

For soils with cohesion and with $\phi' > 0°$ a general solution [3] has been derived based on the friction circle, or ϕ-circle, method. Stability numbers for use in Equation 11.10 are provided in Table 11.1. In all cases in soils with $\phi' > 0°$ (except for $\phi' < 3°$) the critical circle passes through the toe of the slope. In Table 11.1, when a more dangerous circle passes below the toe, the stability number values are shown in parentheses.

TABLE 11.1 Stability numbers for critical circles for homogeneous slopes by ϕ-circle method [3].

Slope Angle, β	Drained Friction Angle, ϕ	Stability number	Slope Angle, β	Drained Friction Angle, ϕ	Stability number
90	0	3.83	45	0	(5.88)
	5	4.18		5	7.35
	10	4.59		10	9.26
	15	5.03		15	12.05
	20	5.49		20	16.13
	25	6.02		25	22.73
75	0	4.57	30	0	(6.41)
	5	5.13		5	(9.09)
	10	5.78		10	13.33
	15	6.58		15	21.74
	20	7.46		20	40.00
	25	8.55		25	111.11
60	0	5.24	15	0	(6.90)
	5	6.17		5	(14.29)
	10	7.25		10	43.48
	15	8.62			
	20	10.31	All slope angles	0	5.52
	25	12.66			

METHOD OF SLICES

The method of slices was developed in the early 1920s in Sweden and was later refined by Bishop [5] to consider interslice forces to some degree. This analysis method can accommodate complex slope geometries, variable soil layering and strengths, variable pore water pressure conditions, internal reinforcement, and the influence of external boundary loads, *but it is only applicable to circular slip surfaces*. It accomplishes this by dividing a slope into a series of vertical slices for analysis, with limiting equilibrium conditions evaluated for each slice, as shown in Figure 11.6.

Each slice can have different layering, different strength, and different pore water pressure than the other slices. If the condition of equilibrium is satisfied for each slice, then it is considered that the entire mass is in equilibrium. The force system on a single slice is shown in Figure 11.7.

The simplified Bishop method ignores the side forces E and X. Provided that the pore water pressure, u, is accounted for, the calculation, results in a very small error compared to more rigorous methods that take the side forces into account. The factor of safety FS is defined in the same manner as in Equation 11.1 in which the same FS is applied to the cohesion and to the friction angle and indicates the fraction $1/FS$ of the cohesion and friction that are mobilized to resist the driving forces.

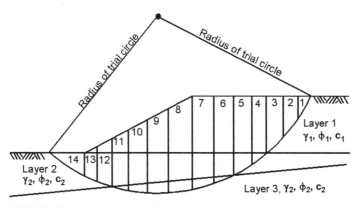

FIGURE 11.6 Typical slope stability analysis using the method of slices.

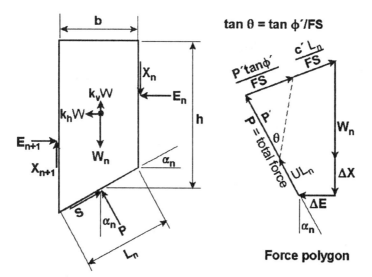

FIGURE 11.7 Forces acting on the nth slice in the Bishop simplified method of slices.

Equation 11.11 gives the factor of safety for a single trial circle. The solution compares the summation of the mobilized resisting forces to the summation of the driving forces, which is the factor of safety for that circle. Because the *FS* term is on both sides of the equation, it is necessary that different values of *FS* be tried until one is found that satisfies the equation. The convergence is usually very rapid. The identical *FS* is applied to all slices in a trial circle and compared in the summation as shown in Equation 11.11.

$$FS = \frac{\sum \left[\left\{ c'b + \left[W\left(1 - \frac{u}{\gamma h}\right)\right] \tan \phi' \right\} \frac{\sec \alpha}{m_\alpha} \right]}{\sum W \sin \alpha} \qquad (11.11)$$

where

$$m_\alpha = 1 + \frac{\tan \phi' \tan \alpha}{FS} \qquad (11.12)$$

The factor of safety for a single trial circle using Equation 11.11 depends on the location of the center of the circle and on the radius of the circle as shown in Figure 11.6. The calculations must be repeated for a series of trial circles with different centers and different radii until the minimum value of *FS* is found. A common minimum *FS* is 1.5.

A minimum of eight slices should be used so that the use of a straight line to approximate the curved lower part of each curve will not lead to excessive error. The scale drawing and calculation procedure is tedious and time consuming if done by hand. Several programs that quickly compute the minimum *FS* are available commercially for use on personal computers.

SEISMIC SLOPE STABILITY ANALYSIS

Earthquake effects can be introduced into an analysis as a pseudostatic force by assigning a disturbing force on the sliding mass equal to *kW* where *W* is the weight of the sliding mass and *k* is the seismic coefficient. Figure 11.7 indicates the seismic coefficient resolved into horizontal and vertical components, k_h and k_v.

Kramer [6] cautions that the use of a pseudostatic force, as in Equation 11.12, has significant shortcomings because the effect of earthquake shaking is a complex and transient phenomenon. Pseudostatic analyses can be unreliable for soils that build up large pore water pressures or show more than about 15 percent degradation of strength due to earthquake shaking [6]. In spite of its limitations, a pseudostatic approach has many attractive features, including its similarity to static stability analysis. Newer, more realistic methods of analysis based on the evaluation of permanent slope deformation are increasingly being used in place of pseudostatic analyses.

EXAMPLE PROBLEMS

EXAMPLE 11.1
The slope shown in Figure 11.8 consists of a sand layer (Layer 1) over a clay layer (Layer 2). What is the factor of safety for the trial analysis for Wedge 3 located as shown and having a length of 52 ft?

SLOPE STABILITY **11.11**

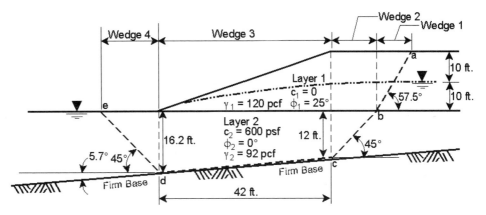

FIGURE 11.8 Example of stability analysis of translation (sliding wedge) failure, to accompany Problem 11.1 [1].

Solution:
The term for factor of safety appears on both sides of the equations for active forces and passive forces. Therefore, it is necessary to solve for $\Sigma P\alpha = \Sigma P\beta$ using various values of factor of safety until equality is reached.

WEDGE 1: $\phi = 25°, c = 0$ psf, $\gamma = 120$ pcf. (sliding surface $a\ b$)
$\alpha_1 = 45 + \phi_1/2 = 57.5°$
$W = 20/2\ (\ 20\ \tan 32.5°\ (\ 120 = 15{,}290$ lb
$P_w = (0 + 10)/2\ (\ 62.4\ (\ (10/\sin 57.5°) = 3680$ lb

$$P\alpha_1 = (W - P_w \cos \alpha_1) \frac{\tan \alpha_1 - \dfrac{\tan \phi_1}{FS}}{1 + \tan \alpha_1 \dfrac{\tan \phi_1}{FS}} + P_w \sin \alpha_1 \qquad (11.2)$$

$$P\alpha_1 = (15{,}290 - 1980)\ \frac{\left(1.57 - \dfrac{0.47}{FS}\right)}{1 + \dfrac{0.73}{Fs}} + 3110$$

$$P\alpha_1 = \left(\frac{20{,}900 \times FS - 6260}{FS + 0.73}\right) + 3110$$

WEDGE 2: $\phi = 0°, c = 600$ psf, $\gamma = 92$ pcf. (sliding surface $b\ c$)
$\alpha_2 = 45 + \phi_2/2 = 45°$
$W = 12 \times 20 \times 120 + (12 \times 12)/2 \times 92 = 35{,}420$ lb

$$P\alpha_a = W \tan \alpha - \frac{c}{FS} \frac{L}{\cos \alpha} \qquad (11.7a,\ \text{for } \phi = 0°)$$

$$P\alpha_2 = 35{,}420 - \frac{\frac{600}{FS}\left(\frac{12}{0.707}\right)}{0.707} = 35{,}420 - \frac{14{,}400}{FS}$$

WEDGE 3: $\phi = 0°$, $c = 600$ psf, $\gamma = 92$ pcf. (sliding surface $c\,d$)
$\alpha_3 = 5.7°$
$W = 20/2 \times 42 \times 120 + (12 \times 16.2)/2 \times 42 \times 92 = 104{,}880$ lb

$$P\alpha_3 = (104{,}880 \times 0.10) - \left[\frac{\left(\frac{600}{FS}\right)\left(\frac{42}{0.99}\right)}{0.99}\right] = 10{,}490 - \frac{25{,}710}{FS}$$

WEDGE 4: $\phi = 0°$, $c = 600$ psf, $\gamma = 92$ pcf. (sliding surface $d\,e$)
$\beta_1 = 45°$
$W = 16.2/2 \times 16.2 \times 92 = 12{,}070$ lb

$$P\beta_1 = W\tan\beta + \frac{c}{FS}\frac{L}{\cos\beta} \qquad (11.7b,\ \text{for}\ \phi = 0°)$$

$$P\beta_1 = 12{,}070 + \left[\frac{\left(\frac{600}{FS}\right)\left(\frac{16.2}{0.707}\right)}{0.707}\right] = 12{,}070 + \frac{19{,}400}{FS}$$

Equating $P\alpha_1 + P\alpha_2 + P\alpha_3 = P\beta_1$ and trying various values of FS, equality occurs at $FS = 1.27$.

EXAMPLE 11.2

A very long slope consists of a sand layer, having a constant 9.1-ft thickness, over rock. The sand layer has the properties $c' = 0$ psf and $\phi' = 32°$, dry unit weight $\gamma = 120$ pcf and $\gamma_{sat} = 125$ pcf. The slope angle $\beta = 25°$. What is the factor of safety if the sand remains dry?

Solution
For a dry sand, Equation 11.9 governs:

$$FS = \frac{\tan\phi}{\tan\beta} = \frac{0.625}{0.466} = 1.34$$

EXAMPLE 11.3

What is the factor of safety for the sand slope of Example Problem 11.2 if rainfall causes seepage parallel to the slope with a constant water level in the sand at ½ of the thickness?

Solution:
For a sand with a vertical height $h = t/\cos\beta = 9.1/0.906 = 10.0$ ft and a constant water level at $m = 0.5$, Equation 11.8 governs:

$$FS = \frac{c' + h\cos^2\beta[(1-m)\gamma_m + m\gamma']\tan\phi'}{h\sin\beta\cos\beta[(1-m)\gamma_m + m\gamma_{sat}]} \quad (11.8)$$

$$FS + \frac{10\cos^2 25[(1-0.5)120 + 0.5(125 + 62.4)]\tan 32}{10\sin 25\cos 25[(1-0.5)120 + 0.5(125)]}$$

$$FS = \frac{469}{469} = 1.0$$

The slope is on the verge of failure!

EXAMPLE 11.4
The slope shown in Figure 11.9 was cut at angle of $\beta = 40°$ in a homogeneous layer of clay. The clay has properties $c = 650$ psf and $\phi = 0°$. A hard layer exists at a depth of 46 feet from the top of the slope. (a) Estimate the factor of safety of the slope, and (b) Locate the position of the center of the circle.

Solution:
For clay with $c = 650$ psf, $\phi = 0°$, $\gamma = 110$ pcf, a slope angle of $\beta = 40°$, and for $d = D/H = 18/28 = 0.64$, Figure 11.4 indicates a stability number $N_0 = 5.75$ which is a base circle. Using Equation 11.10:

$$FS = N_0 \frac{c}{\gamma H} = \frac{650}{110 \times 28} = 1.21$$

From Figure 11.5, $x = 0.7$ and $y = 1.6$. The x-coordinate of the center of the circle, measured from the toe of the slope, is $0.7 \times 28 = 20$ ft. The y-coordinate of the center of the circle, measured from the toe of the slope, is $1.6 \times 28 = 45$ ft. These distances are shown on Figure 11.9.

EXAMPLE 11.5
Assume that the slope shown in Figure 11.9 consists of a homogeneous layer of clayey silt overlying a hard layer. All dimensions are the same as in Figure 11.9 except that the engineering properties of the clayey silt $c = 150$ psf and $\phi = 20°$. Estimate the factor of safety of the slope.

Solution:
For clayey silt with $c = 150$ psf, $\phi = 20°$, $\gamma = 110$ pcf, a slope angle of $\beta = 40°$, and for $d = D/H = 18/28 = 0.64$, interpolation in Table 11.1 indicates a stability number $N_o = 24.1$. Using Equation 11.10:

$$FS = N_0 \frac{c}{\gamma H} = 24.1 \frac{150}{110 \times 28} = 1.17$$

EXAMPLE 11.6

Referring to Figure 11.7, a slope in a homogeneous soil having a unit weight $\gamma = 120$ pcf, cohesion $c' = 300$ psf and $\phi' = 30°$ is subdivided into 10 slices for a simplified Bishop method of slices analysis. The No. 4 slice has a width $b = 20$ ft, a height $h = 31$ ft, and a base angle $\alpha = 21°$. The depth to a static water table is 19 ft. Assuming (for this example problem only) that the calculation for factor of safety FS for this individual slice is the average for all slices, what is the FS for the slope?

Solution:
Solving Equation 11.11:

$$u = \gamma_w b h_w = 62.4 \times (31 - 19) = 749 \text{ lb/ft}^2$$

$$c'b = 300 \times 20 = 6000 \text{ lb/ft}$$

$$\left[W\left(1 - \frac{u}{\gamma h}\right)\tan \phi'\right] = 20 \times 31 \times 120\left[1 - \frac{749}{120 \times 31}\right]\tan 30 = 34{,}364 \text{ lb/ft}$$

$$m_a = 1 + \frac{\tan\phi' \tan\alpha}{FS} = 1 + \frac{0.222}{FS}$$

$$W \sin \alpha = 74{,}400 \times 0.358 = 26{,}663 \text{ lb/ft}$$

$$FS = \frac{(6000 + 34{,}364) \sec\alpha}{26{,}663} \times \frac{1}{m_a} = \frac{1.622}{m_a}$$

Try $FS = 1.50$. The solution of Equation 11.11 is $FS = 1.42$. Not balanced.
Try $FS = 1.45$. The solution of Equation 11.11 is $FS = 1.41$. Not balanced.
Try $FS = 1.40$. The solution of Equation 11.11 is $FS = 1.40$. Balanced.

REFERENCES

1. NAVFAC DM 7.01. 1986. *Soil Mechanics, Design Manual DM 7.01*. Naval Facilities Engineering Command, Department of the Navy. Alexandria, VA.
2. Abramson, L. W., Lee, T.S., Sharma, S., and Boyce, G. M. 1996. *Slope Stability and Stabilization Methods*. New York: John Wiley & Sons.
3. Taylor, D. W. 1937. "Stability of Earth Slopes," *Journal of the Boston Society of Civil Engineers*, vol. 24, no. 3, pp. 197-246.
4. Duncan, J. M. 1996. "Soil Slope Stability Analysis," Chapter 13, *Landslides: Investigation and Mitigation*, A. K. Turner and R. L. Schuster, editors, Special Report 247. Transportation Research Board, National Academy of Sciences-National Research Council. Washington, DC.
5. Bishop, A. W. 1955. "The Use of the Slip Circle in the Stability Analysis of Slopes," *Gèotechnique*, vol. 5, no. 1, pp. 7-17.
6. Kramer, S. L. 1996. *Geotechnical Earthquake Engineering*. Upper Saddle River, NJ: Prentice-Hall.

CHAPTER 12
COMPACTED FILL

PURPOSE OF COMPACTION

Soil or rock fill is placed at a construction site to raise the grade, level the site, backfill a wall, erect a berm, or create similar earthworks. The soil or rock in the fill can be compacted mechanically, or densified, to increase its shear strength and reduce its compressibility and permeability. When the compaction work complies with an engineer's specifications, the fill can be used to support a structure, such as a building or pavement, or to form a retaining structure such as a berm, a dike, or an earth dam. The factors that influence the specifications for an engineered fill are:

1. Raw materials. What are the relevant properties of the available raw materials? How variable are their material properties (see Chapter 2) in the borrow area?

2. End product. What quality is wanted in the end product (the fill)? That is, what engineering characteristics is the compacted soil fill required to have?

3. Compaction process. How should the raw materials be processed to achieve the required end-product properties? What kind of equipment should be used? Are there any processing limitations? Lift thickness? Water content? Should the properties of the raw materials be modified before use?

4. Quality assurance. How can we assure the owner and ourselves that the contractor probably has achieved the specified quality? What should we require the contractor to do for quality control and what should we do for quality verification?

5. Specifications. How can we specify the acceptable or desired materials and compaction process so an earthwork contractor can bid for the compacted fill work realistically and complete the work successfully?

The most common way to densify soils is by mechanical methods, using special rollers or vibrators. Alternatively, the wheels or tracks of heavy hauling equipment can be used.

Mechanical compaction can be done either on hauled or dredged fill soils, one layer at a time; on the surface layer of existing soils; or throughout a layer of clean sand by vibration.

Mechanical densification, or decrease in porosity, is the result of the expulsion of air from the void spaces between the grains. Except in clean gravels and coarse sands, water is not expelled from the soil mass during densification because of the soil's low permeability and the brief time the mechanical force is applied. The two major soil type groups, *cohesionless* and *cohesive*, react differently to the compaction process and will be discussed separately below.

BASIC CONCEPTS

Zero Air Voids Curve

It is possible, with the weight-volume relationships presented in Chapter 2, to construct a graph of dry density versus water content on which a line connects all points of 100 percent saturation, called the *zero air voids curve*. Examples of the simple calculations needed for constructing the curve are presented in Problem 12.1, and curves for several values of specific gravity are shown on the figure accompanying that problem.

The zero air voids curve is important for evaluating compaction test data. It is impossible to have a dry density-moisture content point to the right of the curve for the correct specific gravity because that would imply a degree of saturation greater than 100 percent. Whenever a point plots to the right of the zero air voids curve, either the chosen specific gravity line is too low or there is an error in the test data.

Pressure Distribution with Depth

Mechanical compaction uses various rollers to compress a layer of soil. The distribution of a pressure on the surface of an elastic mass with depth was discussed in Chapter 4. The theoretical solution developed by Boussinesq assumes a uniform contact with the surface. Although this solution is not completely valid because soil is not elastic, the theoretical values are reasonable approximations of the true pressure distribution under a roller.

Figure 12.1 shows an example of the manner in which pressure distributes and diminishes with depth under the wheel of a rubber-tired roller. When the depth is stated in terms of a fraction of the diameter of the equivalent circle (z/d), the three curves lie on top of each other, consistent with the pressure distribution for a uniformly loaded circle shown in Figure 4.13.

COMPACTION OF COHESIONLESS SOILS

Cohesionless soils are mainly gravels, clean sands, and non-plastic silt. They generally contain 5 percent or less by weight of material finer than the No. 200 screen (0.074 mm) and are free flowing. They have no measurable plasticity because of a very low clay content and do not exhibit cohesion in shear.

Clean granular soils that have been removed from a borrow area do not exist in the honeycomb structure of Figure 6.8, but only in a simple loose to medium-dense grain-to-grain

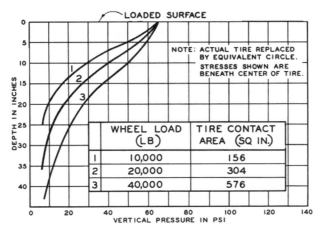

PRESSURE DISTRIBUTION BENEATH WHEEL LOAD
CONTACT PRESSURE OF 65 PSI

FIGURE 12.1 Theoretical distribution of pressure under a rubber-tired wheel [1].

structure. Direct pressure by a weighted roller will not cause appreciable compression because an increase in normal load increases the shear strength, which resists further densification. Only vibrating the cohesionless soil by exerting a lateral shear force on grains without increasing the normal force will cause a decrease in volume, as shown in Figure 7.7.

Factors Affecting Field Compaction of Cohesionless Soils

The relative density achieved during compaction of a clean cohesionless soil is most affected by the following factors:

1. Soil gradation, including the coefficient of uniformity; that is, the shape of the gradational curve, the particle size range, and the particle shape, whether angular or rounded
2. Water content
3. Magnitude of the compactive effort
4. Thickness of the soil layer being compacted
5. Characteristics of the compaction equipment.

Effect of Soil Gradation

The results of a large number of laboratory tests for maximum and minimum density (Chapter 5) were used to develop the relationship shown in Figure 12.2. Note the small difference of about 20 to 25 lbs/ft^3, between the maximum density and minimum density at any given coefficient of uniformity. Similar relationships with maximum grain size and grain angularity can be observed.

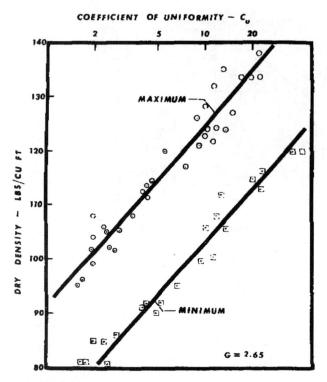

FIGURE 12.2 Maximum and minimum densities from laboratory tests related to the coefficient of uniformity for a clean sand [1].

Effect of Water Content

Vibration-induced densification occurs best on completely dry or completely saturated soils. Partial saturation causes a negative pore water pressure, or suction, in the voids. As sands are wetted, the apparent shear strength decreases and greater densification can be achieved for the same compactive effort.

Effect of Lift Thickness and Magnitude of Effort

Vibratory compactive effort acts throughout the upper part of the soil mass, causing a lifting and lateral motion of the soils nearest the surface. Therefore, maximum densification occurs far enough below the surface so that there is sufficient weight to hold the particles in place, yet close enough to the surface so that the vibratory energy is not appreciably dissipated. This effect is shown in Figure 12.3a, which shows the result of a series of field density tests made in a sand fill that was compacted by vibration.

The uppermost 12 inches (30 cm), more or less, of the soil is less densified than the material immediately below. Thus, placing the sand in relatively thin lifts causes the upper portion of the previous lift to become the lower portion of the new lift, and to become the well-compacted segment of the profile. This is shown in Figure 12.3b.

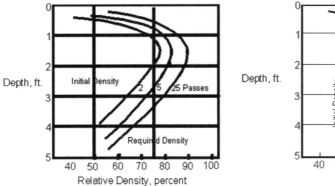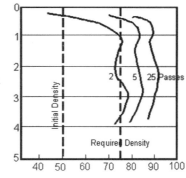

a. Compaction with 6-ton roller on surface of sand layer 6 feet thick

b. Compaction with 6-ton roller on several lifts, each 1.5 feet thick

FIGURE 12.3 Effect of lift thickness and compactive effort on the densification of a cohesionless soil [2].

Effect of Characteristics of the Compaction Equipment

The effectiveness of vibratory compaction depends on several factors involving the interaction between the equipment and the soil being compacted. The factors include:

1. **Operating frequency.** This must be above resonant frequency. The lighter the vibrator, the greater the frequency needed to achieve densification.
2. **Relation of dynamic force to static weight.** Dynamic force should be two to three times greater than the static force of the roller. For vibrating base-plate compactors, the dynamic force should be eight to ten times greater than the static force.
3. **Amplitude of vibration displacement.** Greater amplitude provides greater force but requires greater power.
4. **Speed of travel.** A vibratory roller should travel about 1 to 2 mph. A vibrating base plate compactor should be moved about 0.5 to 1 mph.
5. **Properties of soil to be compacted.** Vibratory compactors normally are used only for clean granular (cohesionless) soils, but the amplitude provides an effective tamping action on some low plasticity cohesive (clay, mixed-grain) soils.

Deep Compaction by Vibration. Two proprietary methods have been developed to compact by vibration sand deposits having depths greater than 60 ft (20 m) without removal and replacement of the sand in layers.

Vibroflotation. Vibroflotation uses a torpedo-like vibratory unit, called a vibroflot, which is jetted into the ground. A typical probe has a diameter of about 16 in. (40 cm) and a length of about 7 ft (210 cm), weighs about 4000 lbs (2 tonne) and vibrates about 30 Hz. The system is operated in the following manner:

1. The vibroflot unit is suspended from a crane.

2. Using a centerline water jet, the unit is lowered into the sand deposit under its own weight.
3. When the desired depth is reached, vibration is started while the water flow is maintained. The upward water flow around the outside of the vibroflot keeps a channel open along the side for backfill as sand is added at the surface to make up the volume lost due to densification.
4. The probe is raised in increments, compacting the surrounding sand, and more backfill is continually added until the probe reaches the surface.

Vibrating Pile Probes. The L. B. Foster Company Terra Probe is an open-ended steel-pipe pile attached to a vibratory pile driver-extractor. The pipe has a diameter of 30 in. (75 cm) and a length of 45 ft (13.7 m), weighs about 20,000 lbs (10 tonne), and vibrates about 15 Hz. The compaction operation involves vibratory driving of the pile, without the use of water jetting, and slow withdrawal of the vibrating pile, causing a compacted sand zone to be formed around the pile.

Determining Effectiveness. Typically, the vibrating probes are installed in a pattern so that the circles of zones of compaction somewhat overlap. Therefore, the location of least compaction is at the intersection of the circles, farthest from any of the probes.

The effectiveness of the specified vibratory compaction program should be verified, either to confirm whether the selected pattern and probe spacing is valid or needs modification, or to test the quality of the contractor's work, or both. The depth of vibratory compaction does not normally permit field density testing except in shallow pits. Therefore, it is common to use a soil-drilling rig to perform either standard penetration tests or cone penetration tests (see Chapter 7), normally at the center point between a group of probes.

COMPACTION OF COHESIVE SOILS

The compaction of cohesive soils occurs by dynamically compressing the soil by forcing air out of the voids. To accomplish this, the shear strength of the soil mass must be overcome in a manner similar to that realized by an extremely rapid consolidation test (Chapter 4). The densification of cohesive soils is affected by several factors, including:

1. Soil texture, including clay content and plasticity
2. Water content and degree of saturation
3. Magnitude of compactive effort
4. Thickness of soil layer (lift)
5. Characteristics of the compaction equipment.

Compactive effort is defined as the work input of the compaction equipment, measured in foot-pounds per cubic foot of soil. The work input depends on the size and type of equipment, the number of passes, and the thickness of the layer to be compacted. In the field, it often is measured in terms of drawbar pull times the lift thickness times the number of passes.

Example of a Test Fill

Assume that a test fill was made in the pattern shown in Figure 12.4.

Table 12.1 contains the measurements of density and water content after compaction with a heavy sheepsfoot roller of several sections of a silty clay soil with various water contents. (NOTE: These data are for illustration purposes only. They were not derived from actual field measurements).

The results of the field density tests are shown in Figure 12.5.

Figure 12.5 illustrates that the compaction of cohesive soils should be viewed as a three-dimensional relationship. No real *maximum density* or *optimum water content* exists for any given soil because dry density is affected by both the moisture content and the

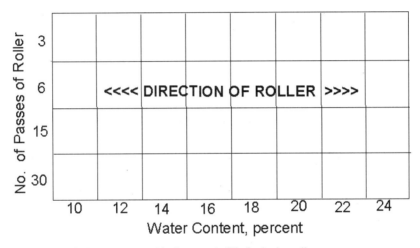

FIGURE 12.4 Test pattern used in the example fill of cohesive soil.

TABLE 12.1 Data for Example Test Fill

No. of Passes * w %	3 PCF	6 PCF	15 PCF	30 PCF
10	90.0	95.5	103.0	108.5
12	92.0	98.5	106.0	113.5
14	94.5	102.0	110.0	116.5
16	97.0	104.7	111.5	114.5
18	99.0	104.9	108.8	111.0
20	100.5	103.0	105.7	107.1
22	98.7	100.6	101.3	103.8
24	97.3	98.0	99.2	100.6

* No. of Passes = Number of uniform passes of a heavy sheepsfoot roller on a test fill of silty clay soil with a loose layer thickness of about 9 inches (23 cm) and a compacted thickness of about 6 inches (15 cm).

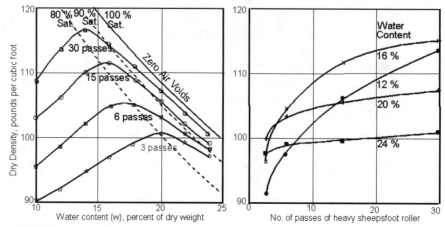

EXAMPLE OF FIELD DENSITY TESTS MADE ON A TEST FILL OF SILTY CLAY COMPACTED WITH A HEAVY SHEEPSFOOT ROLLER.

FIGURE 12.5 Results of field density tests made on example test fill.

compactive effort. The combination effect shows that at a low moisture content (12 percent), the compactive effort must overcome the shear strength due to a high negative pore water pressure. Density increases with the number of passes at a decreasing rate and the soil becomes stronger with density and as the zero air voids curve (saturation) is approached.

As the moisture content is increased (16 percent), the effectiveness of the compactive effort increases because the shear strength is less because of a lesser suction. Again, density increases at a slower rate as the zero air voids curve is approached.

When the moisture content reaches a level at which the soil is very wet (20 percent) and as the density approaches the zero air voids curve, the compactive energy is spent in simply compressing the water in the voids rather than shearing and compressing the soil.

Finally, at a high enough water content (24 percent), virtually all of the compactive energy is spent in simply compressing the water in the voids rather than shearing and compressing the soil, and no increase in density is achieved by further rolling.

The left side of Figure 12.5 indicates that the maximum density for any given compactive effort is achieved when the degree of saturation reaches between 80 percent and 90 percent. This has been confirmed in a large number of field density tests made at project sites [3].

Effect of Soil Texture

Because of the effect of partial saturation on the finest fraction of a cohesive soil, soils containing an appreciable amount of coarse grains are easier to compact to higher densities. Figure 12.6 shows the effect of soil texture, that is, grain-size distribution, on maximum density and optimum water content for a constant compactive effort. The coarsest materials have the highest maximum density and, as the clay content increases, the maximum density decreases.

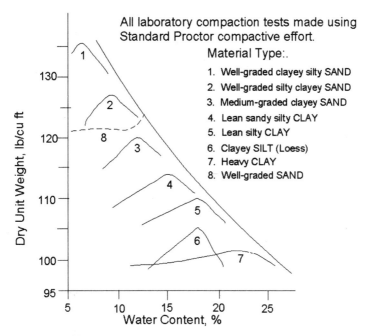

FIGURE 12.6 Variation in maximum density and optimum water content using Standard Proctor compactive effort for soils of several textures [4].

Effect of Lift Thickness

As discussed above, the pressure under a roller decreases with depth within the layer being compacted. This is illustrated by Figure 12.7, which shows the results of a test embankment prepared at the USAE Waterways Experiment Station. A test plot was made using a single rubber-tired roller on areas with various water contents and either 6-, 12-, or 24-inch lifts. About the same density was obtained in the top 6 inches regardless of total lift thickness. For thick lifts, the density, however, decreased rapidly with depth.

Effect of Field Compaction Equipment

Virtually all compaction equipment for cohesive soils uses some form of kneading—the same pushing, pressing, and releasing action used in kneading bread dough. The action causes a mixing of the soil and the expulsion of air from the voids by compression of the loose soil structure.

Effect of Compacting Moisture on Engineering Properties

The water content and compactive effort at which a given soil should be compacted to achieve a desired shear strength and/or permeability depends on the suitability of the soil; the ease of modifying the existing water content, particularly in an area of high rainfall; and the availability of compaction equipment of a given weight.

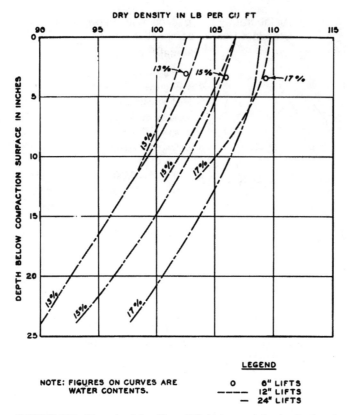

FIGURE 12.7 Example of the effect of lift thickness on dry density, based on field test data [2].

The maximum compaction efficiency occurs when the soil is compacted at a water content that will reach from about 80 percent to 90 percent saturation at the desired density. Further application of compactive effort produces only a decreasing rate of densification because of increased saturation. No additional saturation appears to be possible when the degree of saturation reaches from about 95 percent to 97 percent. Because compaction imparts a preconsolidation pressure to the soil, the greater the densification the lesser the compressibility of the soil.

Strength of Cohesive Soil. At any specific density and water content, a compacted cohesive soil will have a shear strength that depends, in large part, on the surface tension (negative pore water pressure)—a function of the degree of saturation. The example soil shown in Figure 12.8 is the same as that shown earlier in Figure 12.5 for the data in Table 12.1.

The effect of saturation on a cohesive soil placed dry of *optimum* is shown in Figure 12.8. Assume that a soil is compacted at a low moisture content, for example 12 percent, as shown by the dashed line in the upper part of the figure. Then the compacted fill is inundated, perhaps because it is in the interior of an earth dam or because of intense rainfall, and the degree of saturation rises at a constant density to about 24 percent moisture. The shear strength of the compacted soil drops, in the example shown, to about one-half its

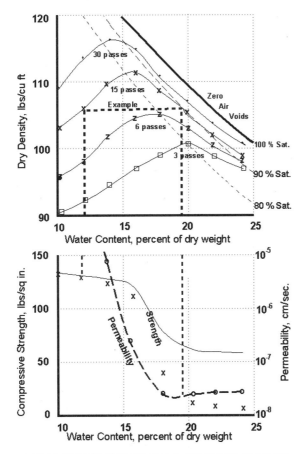

COMPACTION TESTS OF A SILTY CLAY SOIL

FIGURE 12.8 Effect of compaction on engineering properties of soil.

original strength. The designer must consider this possibility in specifying the amount of densification that must be achieved in the field.

Permeability of Cohesive Soil. Clay particles exist mainly as flakes or platelets rather than in a roughly spherical shape. Because of electrical charges imparted by ions in the pore water, some clays exist in nature with a highly disorganized, or *flocculated*, structure, in which the platelets have an end-to-flat, random orientation, as shown in Figure 12.9. In a *dispersed* structure, the clay flakes are all roughly parallel and can exist closer together than they can in the flocculated structure.

If a flocculated clay is compacted at a high degree of saturation, near the zero air voids curve, the flocculant structure can be destroyed and the dispersed structure will be developed. Clays having a dispersed structure exhibit lower shrinkage, lower swelling, much lower permeability, and lower strengths at low strains than flocculated samples of the same density and water content. The lower permeability is shown for the example in Figure 12.8.

Highly Flocculant Structure | Moderately Flocculant Structure | Dispersed Structure

FIGURE 12.9 Clay platelet structure of cohesive soils.

The flocculated structure of Figure 12.9 occurs in soils that have been deposited out of suspension in to sea water or ion-bearing fresh water. This includes nearly all sedimentary deposits of clay. The dispersed structure is typical of mixed or remolded soils, whether natural or manmade. Natural sources include glacial till and soils sedimented in the presence of a dispersing agent.

Modifying Compactibility with Chemicals

Common chemicals mixed with clayey soil will modify the compactibility of the soil by either reducing the surface tension resulting from partial saturation, or changing the clay mineral formulation by the addition of lime. These chemicals are useful only in subgrade soils containing an appreciable amount of clay. Surfactants, primarily sodium phosphates (soap), can be used to reduce surface tension in the water in clayey soils, thereby assisting in the dispersion of the clay particles. In contrast, the addition of lime causes flocculation of the clay particles. Lime also replaces the potassium and sodium ions with calcium, reducing the plasticity index. It is one of the most cost-effective methods available for permitting compaction of high plasticity clays by reducing the needed compactive effort.

QUALITY ASSURANCE TESTING

Quality Assurance Definitions

The objective of all tests and observations made during construction of a compacted fill is to provide a reasonable assurance that the quality of the completed fill is as specified. This is typically achieved by some form of quality assurance program.

Quality assurance (QA) consists of all the planned and systematic actions necessary to provide adequate assurance to the owner that a structure or system (in this case, a compacted fill) will perform satisfactorily in service; that is, in accordance with the design and specification requirements. Quality assurance programs generally include both quality control procedures undertaken by the contractor and quality verification programs undertaken by the owner. These terms are defined below because some engineers use the term *quality control* when, in fact, they mean *quality verification*.

Contractor Quality Control (CQC) consists of the contractor's measurements or observations of construction quality characteristics. The contractor's oversight is significant for controlling the materials and workmanship to meet specification requirements, combined with a system for initiating corrective action when off-standard quality is encountered. Only the contractor has control over his or her personnel, equipment, materials, and workmanship. If the project specifications permit, however, the owner can require that the contractor perform certain quality control test procedures and report to the owner's representative as part of the quality verification program.

Quality Verification (QV) is the owner's action of sampling and testing or observing a block of material or workmanship the contractor has presented as ready, for comparison with the specification requirements and which result in acceptance or rejection of the block. The owner, through his or her engineering and testing technician representatives, can only accept or reject the materials or workmanship presented by the contractor for evaluation. Attempts by the owner's personnel to control the contractor's operation usually lead to claims and are regarded in the courts as an assumption of responsibility for the quality of the resulting work.

The QV tests can be the same as the contractor's CQC tests, and often the contractor uses the owner's QV tests as part of his or her CQC program. The distinction between CQC and QV tests, however, must be maintained because their objectives are different. The contractor uses CQC tests to prepare the work for acceptance by the owner and the owner uses QV tests to evaluate the work for acceptance or rejection.

The Proctor Compaction Test for Cohesive Soil

In the late 1920s and early 1930s, it was common for compaction specifications to require a given type and size of roller, lift thickness, and number of passes. Test fills, using the multiple water content and multiple strip procedure described in the example above, were used to determine the optimum water content for the contractor's specific equipment and desired number of passes for a given soil.

In 1933, Raymond R. Proctor published a series of four articles [5] describing the relationships shown above for the test fill. Proctor described a laboratory test procedure that could be used to define the moisture-density relationship for a given soil and would provide the same optimum water content as that obtained from the very expensive test fill procedure.

Two of Proctor's concerns were the effect of near-saturation *after* compaction on the strength of a compacted soil, and trying to convince contractors to compact soils at *optimum moisture*. Another purpose for the laboratory compaction test was to provide samples for laboratory strength tests so that the field equipment and compactive effort could be selected without obtaining field samples, a comparatively expensive and time-consuming practice [6].

Later minor modifications were made to Proctor's test device, resulting in the now *standard Proctor compaction test*. Because the maximum density and optimum water content are sensitive to soil composition, as shown in Figure 12.8, the Proctor compaction test procedure was adopted in the late 1930s and early 1940s as a specification standard for quality assurance testing of fill compaction. This is a radical departure from Proctor's original purpose of modifying the test procedure to fit the compactive effort of the equipment and methods being used in the field.

The fact that heavy rollers used in base course construction achieve densities greater than the maximum from the standard Proctor test, caused some people to question the test method. Therefore, a method with greater compactive effort, called the *modified Proctor test*, was developed. The two tests are summarized in Table 12.2.

Field Density and Water Content Test Methods

As part of a quality control or verification program, the density and water content of each layer of compacted fill should be measured as construction proceeds to determine the effectiveness of the compaction method and the contractor's compliance with specifications.

Field Density Test Methods. There are two basic procedures, as described in Table 12.3, for determining the in situ density of a soil mass, *direct measurement* and *indirect measurement*.

The direct field excavation methods are based on the procedure of digging a hole, measuring its volume, and weighing the excavated material. Direct measurement methods include the sand cone method, the water balloon method, and the drive cylinder method. The indirect measurement methods involve passing some form of energy through the soil and measuring the attenuation of the energy as a function of density. The most popular of these is the nuclear gauge method, which uses radioactive energy transmission through the soil, as shown in Figure 12.10.

Water Content Test Methods. Two basic procedures, as described in Table 12.3, for determining the moisture of a soil sample are *direct measurement* and *indirect measurement*. In the direct measurement methods, the wet sample is weighed, the water is removed, and then the dry soil is weighed. The difference in weight is the weight of water in the soil. Methods of drying include using an oven at 105°C, a field hot-plate (not recommended, but used), a microwave oven, and a calcium carbide (Speedy Tester) device. The indirect measurement method, the nuclear gauge of Figure 12.10, uses the thermalization (slowing down) of neutrons that are affecting hydrogen ions. The results often require several hours for oven drying but are available within a few minutes with a calcium carbide tester or a nuclear gauge.

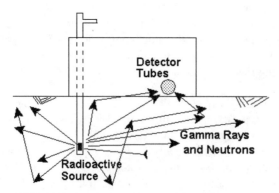

FIGURE 12.10 Nuclear moisture-density gauge.

TABLE 12.2 Proctor compaction test methods

Test Method	ASTM Standard [7]	Equipment and Procedure
Standard	D 698	Mold is 1/30 ft^3 (4-in. diameter). Soil is placed in 3 layers, with 25 blows per layer of a 5.5-lb hammer with a 2-in. diameter face, falling freely 12 inches. Only soil passing the No. 4 screen (1/4 in.) is used. This yields 12,375 ft-lbs of energy per cubic foot of soil.
Modified	D 1557	Mold is 1/30 ft^3 (4-in. diameter). Soil is placed in 5 layers, with 25 blows per layer of a 10-lb hammer with a 2-in. diameter face, falling freely 18 inches. This yields 56,250 ft-lbs of energy per cubic foot of soil.

TABLE 12.3 Summary of Field Density and Water Content Test Methods

Test Method	ASTM Method	Discussion
Field Density Test Methods		
Undisturbed tube sample	D 2937	Uses a short section of thin-wall sampling tube. Driven into cohesive fill and retrieved. Weight of soil in tube is determined.
Sand cone	D 1556	A small hole, 15 cm (6 in.) deep is excavated. The soil is carefully retrieved and is weighed. The volume of the hole is measured by filling with sand from a container having a cone-shaped end. The weight of calibrated sand is used to calculate volume.
Water balloon	D 2167	A small hole, 15 cm (6 in.) deep is excavated. The soil is retrieved carefully and weighed. The volume of the hole is measured by inserting a rubber balloon in the hole and then filling the hole with water from a calibrated container.
Nuclear density gauge	D 2922	A radioactive source of gamma rays is placed in direct contact with the soil. A Geiger counter a short distance away measures the intensity of gamma rays passing through the soil. The energy intensity is proportional to the bulk density. A calibration with materials of known density is used to calculate the field density.
Rapid Water Content Test Methods		
Microwave oven	D 4643	A computer-controlled standard microwave oven is used to dry a soil sample to constant weight.
Calcium carbide gas pressure	D 4944	Water in a soil sample combines with calcium carbide in a closed container. The pressure of the cetylene gas is proportional to the water content.
Rapid heating	D 4959	Soil sample may be dried in a pan on a hot plate or burner. Temperature is not controlled; therefore a slight error may be introduced. Useful for coarse-grained soils .
Nuclear moisture gauge	D 2922	The radioactive source of the nuclear density gauge also emits neutrons. When the neutrons strike a hydrogen atom, the velocity is halved (thermalized). A counter measures only the fast neutrons. The reduction is proportional to the water content.

Laboratory and Field California Bearing Ratio Test

During 1928 and 1929, the California Division of Highways made an intensive investigation of pavement failure throughout the state. A simple and quick laboratory test was sought that would evaluate the quality of subgrade, subbase, and base course materials and establish the density to which they should be compacted in the field. Initially, a static load was used to compact the soil specimens in a laboratory mold. After four days of soaking under a surcharge load, the specimen was tested, in the mold, by penetrating it with a 3 in^2 (6.45 cm^2) steel piston and measuring the resistance at 0.1 inch (0.25 cm) penetration. This resistance then was compared to a standard resistance obtainable from the penetration of compacted, crushed stone. The ratio of the soil resistance to the standard resistance, expressed as a percentage, was called the California Bearing Ratio (CBR).

After Proctor's work in the early 1930s led the Corps of Engineers to adopt the Proctor impact compaction test during the wartime airfield construction of the 1940s, the CBR test was adopted for the design of flexible pavements. The test procedure also was adapted so that the CBR test could be made in the field on existing subgrade soils.

Laboratory CBR Test. The Corps of Engineers modified the CBR test to employ a Proctor-type impact compaction instead of a static load compression. After samples were tested for edge effects due to the size of the piston, it was decided that a larger mold would be used. The standard and modified Proctor tests were adapted to the larger 6-inch (15.2 cm) diameter mold by increasing the number of blows per layer to yield the same compaction energy per cubic foot of soil. For the equivalent to the modified Proctor compactive effort, the test is made using 55 blows per layer instead of 25.

Following ASTM D1883 [7], the laboratory CBR is made as follows:

1. Several conventional modified Proctor compaction tests are made to closely approximate the optimum moisture content.
2. Three specimens are made in the 6-inch diameter CBR mold, each using the modified Proctor procedure, with 55, 25, and 10 blows per layer on 5 layers with a 10 lb (4.54 kg) hammer falling 18 inches (46 cm).
3. A weight not less than 10 lb (4.54 kg) is placed on the compacted soil in the mold to equal the pavement and base-course weight. The specimen in the mold is immersed in water and allowed to swell for four days. The weight is removed and the sample is weighed.
4. A penetration surcharge weight, the weight of which is equal to the pavement and base-course weight, but is not less than 10 lb (4.54 kg), and which has a slot in the center, is placed on the surface of the soil in the mold.
5. A steel penetration piston with a 1.98 in. (5.03 cm) diameter is pressed into the surface of the specimen at the rate of 0.05 inches (0.13 cm) per minute. Load readings are taken at equal intervals of penetration to 0.5 inches (1.3 cm).
6. After correcting the stress-penetration curve to pass through zero, the pressures at 0.1 and 0.2 inches (0.25 and 0.5 cm) are determined. The pressure at 0.1 inches is divided by the standard pressure of 1,000 lbs/in^2 and the pressure at 0.2 inches is divided by the standard pressure of 1,500 lbs/in^2 and converted to percentages. The CBR usually is selected at 0.1 in. penetration. If the CBR at 0.2 in. is greater than at

0.1 in., the test is rerun. If the check test gives similar results, the CBR at 0.2 in. penetration is used.

7. For design purposes, the CBR values of the three tests are plotted so that the CBR at a given percentage (for example, 90 percent or 95 percent) of maximum density can be determined. This is then the CBR value used in pavement design.

Field CBR Test. The CBR of an existing subgrade soil can be field tested by using a similar procedure that is, by covering the cleared and leveled soil with a standard laboratory surcharge equal to the pavement and base course weight, but not less than 10 lb (4.54 kg). A standard CBR penetration piston is pressed into the soil surface at the same rate and the CBR is determined by the same criteria as those used in the laboratory test. If desired, the CBR of the uppermost six inches or more of the existing subgrade can be improved by passing a compaction roller over the surface. The CBR is then retested until the target value is reached.

SPECIFYING SOIL COMPACTION

The fill to occupy a specified space can consist of:

a. *Uncompacted*, or simply dumped and shaped
b. *Semicompacted*, in which lift thickness is unspecified, the soil is placed at natural water content, and compaction is done by uniform coverage of the wheels or tracks of hauling equipment
c. *Compacted*, involving specification of lift thickness, type of compaction equipment, and water content to achieve a specified density.

The following discussion deals only with specifications for semicompaction and compaction. Specifications for compacted fill usually will contain requirements for the following:

1. The soil type, or types, and the sources that can be used and the sources to be excluded
2. The amount of compactive effort or the degree of compaction to be achieved, based on either evaluation of a test fill or a minimum relative density for granular soils or a minimum percentage of the maximum density from a laboratory Proctor compaction test for cohesive soils
3. The type of compaction equipment that may, or must, be used
4. The maximum lift thickness for the selected soil type and equipment.

Acceptable Soil Types

Virtually all types of soil can be compacted successfully. Those that are difficult to compact include high plasticity clays and soils containing large quantities of rock fragments of cobble or larger size. High plasticity clays form very hard clods, although the plasticity can be modified as discussed above, if cost-effective. Large rock fragments can interfere with the compaction of the soil between the fragments.

The presence of intact rock fragments causes the density of the compacted soil to be higher than that used in the laboratory Proctor test. Therefore, it is common to screen and measure the amount of oversize (+¼ in.) material and correct the measured density.

Compactive Effort Requirement

The specifications may prescribe any one of the following methods to achieve the required compaction:

1. A specified number of passes of a roller of a specified or acceptable size, type, or pressure is made uniformly over each layer of a specified maximum thickness. A full-time inspector is required to assure the owner that the requirements are actually met.
2. Repeated passes of a roller of a specified or acceptable size, type, or pressure are made uniformly over each layer of a specified maximum thickness until a specified action occurs, such as deflection under the roller or a limiting penetration of feet of sheepsfoot roller. Again, a full-time inspector is needed to assure the owner that the requirements are actually have been met.
3. Repeated passes of a roller, of a specified or acceptable type, are made uniformly over each layer of a specified maximum thickness until the soil is compacted to a specified density. The density normally is based on a standard laboratory compaction test. The intensity of field density testing is determined by the size and sensitivity of the project.

Cohesionless Soils. Specifications for cohesionless soil compaction usually require that the soil be densified to at least a stated relative density. Typical requirements are for the relative density to exceed from 60 percent to 80 percent. In view of the small range between maximum and minimum densities, it is increasingly common, however, to specify relative compaction as a percentage of the maximum density.

Cohesive Soils The compactive effort requirement for cohesive soils is sometimes stated by specifying the type, size, and weight of the compaction equipment to be used and the minimum number of passes of the roller. More commonly, however, a required density is specified, based on the results of a laboratory Proctor compaction test of the selected soil. Sometimes, water content limits also are placed.

Types of Compaction Equipment

Cohesionless Soils. Equipment for vibratory compaction of cohesionless soils consists of rollers or plates and can be self-propelled or manually propelled. Commonly used types are vibrating rollers, either towed or self-propelled; self-propelled vibrating plates or shoes; and manually propelled vibrating plates.

Cohesive Soils. Compaction equipment for cohesive soils consists of various types of rollers. Various designs of impact tampers are used in space-restricted areas. The most common equipment types for cohesive soils are tamping rollers (sheepsfoot), either towed or self-propelled; smooth-wheel rollers; pneumatic-tired (rubber-tired) rollers; and hand-held tampers (jumping jacks).

A number of compactors are compared in Table 12.4.

Lift Thickness

Cohesionless Soils. Typical lift thicknesses for clean granular soils subjected to vibratory roller compaction are shown in Table 12.5 shown.

TABLE 12.4 Equipment Suitability for Compacting Soils [8]

Type of Compactor	Soil Best Suited For	Maximum Effect in Loose Lift, inches	Density Gained in Lift	Max. Weight Tons
Sheepsfoot	Clay, silty clay, gravel with clay binder	7-12	Nearly Uniform	20
Steel tandem, two-axle	Sandy silts, most granular materials with some clay binder	4-8	Average *	16
Steel tandem three-axle	Sandy silts, most granular materials with some clay binder	4-8	Average *	20
Steel three-wheel	Granular or granular-plastic material	4-8	Average* to uniform	20
Pneumatic small-tire	Sandy silts, sandy clays, gravelly sand, and clays with few fines	4-8	Average* to uniform	12
Pneumatic large-tire	All types	3-6	Uniform	50
Vibratory	Sand, silty sands, silty gravel	up to 24	Uniform	30
Combinations	All	3-6	Uniform	20

* The density can decrease with depth within each lift.

TABLE 12.5 Typical Lift Thickness for Clean Granular Soils

Vibratory Roller Tonne (metric)	Weight Tons	Lift Thickness meters	inches
0.9	1	0.2-0.3	8-12
1.8	2	0.30-0.46	12-18
4.5	5	0.46-0.61	18-24
9.1	10	0.61-0.76	24-30
13.6	15	0.76-1.22	36-48

Cohesive Soils. Lift thicknesses for mechanized rollers used to compact cohesive soils generally are limited to about 23 cm (9 in.) loose to result in a 15 cm (6 in.) thick compacted layer. For steel-wheeled rollers and hand-held tampers, the lift thickness usually is held to no more than from 5 to 8 cm (2-3 in.) thick.

EXAMPLE PROBLEMS

EXAMPLE 12.1
Calculate and plot on a graph the zero air voids curves for specific gravities of 2.60, 2.65, and 2.70.

Solution:
Review the example problems in Chapter 2. For each data point for a given specific gravity, either (a) assume a saturating water content and find the dry density that corresponds, or (b) assume a dry density and find the saturating water content. A spreadsheet, manual or computerized, helps. For example, choose a specific gravity of 2.68. Then for each option,

a. Assume a saturating water content of 14 percent. The unit cube for a volume of one cubic foot is shown in Figure 12.11a.

$$V_s = \frac{W_s}{2.65 \times 62.4} = \frac{W_s}{165.4}$$

$$V_w = \frac{W_w}{62.4} = \frac{0.14 \times W_s}{62.4} = \frac{W_s}{445.7}$$

$$V = V_s + V_w = \frac{W_s}{165.4} + \frac{W_s}{445.7} = 1$$

Solving, $W_s = 120.6$ lbs/ft^3

b. Assume a starting dry density of 115 pcf. The unit cube for a volume of 1 cubic foot is shown in Figure 12.11b.

$$Vs = \frac{115}{2.65 \times 62.4} = 0.695$$

$$V_w = 1 - 0.695 = 0.305$$

$$W_w = 0.305 \times 62.5 = 19.0 \text{ lb}$$

Solving for water content, $w = 16.5\%$

The results of similar calculations for a large number of points for each specific gravity are shown on Figure 12.12.

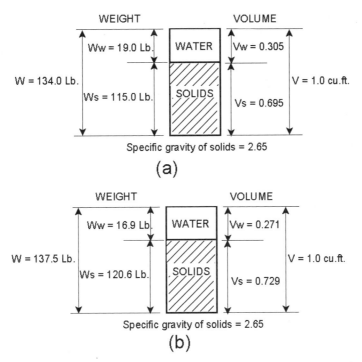

FIGURE 12.11 Unit cubes to accompany calculations for zero air voids curve in Problem 12.1.

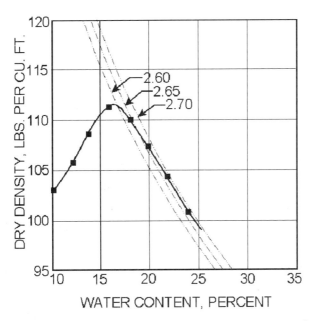

FIGURE 12.12 Moisture-density compaction curve to accompany Problems 12.1, 12.2, and 12.3.

EXAMPLE 12.2

A standard Proctor compaction test was made in the laboratory, using a $1/30\ ft^3$ mold, on a soil assumed to have a specific gravity of 2.65. The results are shown below. Calculate the dry density for all points and plot them on the graph of Problem 12.1.

Wt. of wet soil in mold, lbs	Moisture content, %	Dry density, lbs./ft³
3.78	10	103.0
3.94	12	105.5
4.14	14	109.0
4.31	16	111.4
4.28	20	107.21
4.15	24	100.5

Solution:
The points are plotted on Figure 12.12.

EXAMPLE 12.3

Problem 12.2 uses a specific gravity of 2.65. Is that specific gravity reasonably correct for the soil tested and if not what would be a reasonable value to use? Explain why.

Solution:
No, the three highest moisture content points plot to the right of the zero air voids curve for a specific gravity of 2.65. The correct specific gravity appears to about 2.70 because the high moisture content points are all to the left of the curve and appear to be roughly parallel to the curve.

EXAMPLE 12.4

A field density test was made on a compacted clean sand, indicating a dry density of $114.5\ lbs/ft^3$. The specifications required that the sand be compacted to a minimum of 75 percent relative density. A sample of the sand was tested in the laboratory, resulting in a minimum density of $96\ lbs/ft^3$ and a maximum density of $125\ lbs/ft^3$. (a) Did the compaction meet the specification requirements? (b) Will additional rolling be helpful? Explain.

Solution:

a. Equation 5.10 indicates that the relative density is calculated as

$$D_r, \% = 100\ \frac{\gamma_{d-max}(\gamma_d - \gamma_{d-min})}{\gamma_d(\gamma_{d-max} - \gamma_{d-min})} = 100\ \frac{125(1145 - 96)}{114.5(125 - 96)} = 69.6$$

Therefore, the field compaction did not meet the required 75 percent relative density.

b. As shown in Figure 12.3, continued rolling will be beneficial.

REFERENCES

1. U.S. Army. April 1981. "Construction of Earth and Rock-fill Dams," Lecture Notes, Corps of Engineers Training Course. Geotechnical Laboratory, U.S. Army Engineer Waterways Experiment Station. Vicksburg, MI.
2. D'Appolonia, D. J., Whitman, R. V., and D'Appolonia, E. 1969. "Sand Compaction with Vibratory Rollers." *Journal, Soil Mechanics Division*, American Society of Civil Engineers. 95(SM1).
3. Torrey, V. H. and Donaghe, R. T. 1991. "Compaction Control of Earth-rock Mixtures." Technical Report GL-91-16. USAE Waterways Experiment Station. Vicksburg, MI.
4. Johnson, A. W. and Sallberg, J. R. (1960). "Factors that Influence Field Compaction of Soils: Compaction Characteristics of Field Equipment," Bulletin 272. Highway Research Board, National Academy of Science—National Research Council. Washington, D.C.
5. Proctor, R. R. 1933. "The Design and Construction of Rolled Earth Dams." *Engineering News Record*. 111: 245-248, 286-289, 348-351, 372-376.
6. Proctor, R. R. 1936. "The Relationship between the Foot Pounds per Cubic Foot of Compactive Effort Expended in the Laboratory Compaction of Soils and the Required Compactive Efforts to Secure Similar Results with Sheepsfoot Rollers." *Proceedings of the First International Conference on Soil Mechanics and Foundation Engineering*. Cambridge, MA.
7. ASTM. "Natural Building Stones; Soil and Rock; Geotextiles." American Society for Testing and Materials. Philadelphia, PA.
8. Peurifoy, R.L. and Ledbetter, W.B. 1985. *Construction Planning, Equipment, and Methods*, 4th edition. New York: McGraw-Hill.

INDEX

A

Acceptable soil types, 12.17
Active earth pressure, 10.3
Active lateral pressure due to earthquake forces, 10.7
Activity index, 2.5
Anisotropic soil, flow net in, 3.17
Artesian flow, 3.8-3.9
Atterberg limits, 2.4, 2.18
Atterberg limits and clay content, 2.6

B

Bearing capacity, 8.1
Bearing capacity equations, theoretical, 8.3
Bearing capacity of cohesive soils, 8.8
Bearing capacity of pile groups, 9.12
Borehole shear test, 7.16
Borings, 7.7
Boussinesq equations, 4.10
Braced excavations, design of, 10.18

C

California bearing ratio test, laboratory and field, 12.16
Cohesionless soil, 6.11
Cohesive soil, 6.11, 10.6
Combined stresses, 5.1
Compacting moisture on engineering properties, effect of, 12.9
Compaction, purpose of, 12.1
Compaction of cohesionless soils, 12.2, 12.6
Compaction equipment, effect of characteristics of, 12.5
  Types of, 12.18
Compactive effort requirement, 12.18
Compressibility, 4.1
Compression index, estimating the, 4.8
Consolidated-drained shear of saturated soils, 5.12
Consolidated-drained shear of saturated soils, 5.15
Consolidation, primary, 4.20
  Secondary, 4.23
Consolidation settlement of structures, 4.18
Critical hydraulic gradient, 3.13
Cyclic loading, 8.15

D

Danish formula, 9.14
Darcy's law, 3.1
Deep foundations, types of, 9.1
Deep foundation capacity from load tests, 9.3

Deep foundation capacity from static analysis, 9.7
Deep foundation failure modes, 9.2
Direct shear test, 5.5
Distribution, grain size, 2.1, 2.16
Downward flow, 3.13
Dynamic penetrometer test, 7.20

E

Earthquake loading of foundations, 8.14
Effective stress, 3.11
    Principle of, 3.12
Electrical resistivity surveys, 7.4
Empirical estimates for coefficient of permeability, 3.6
Engineering news formula, 9.14
Estimating drained friction angle, 7.26
Estimation of liquefaction potential, 7.29
Estimating pre-consolidation pressure, 7.27
Estimating relative density and friction angle, 7.23
Estimating unconfined compressive strength, 7.23, 7.25, 8.13
Equivalent horizontal permeability, 3.4
Equivalent permeability of stratified deposits, 3.4
Equivalent vertical permeability, 3.6
Example problems, 2.16, 3.20, 4.24, 5.20, 6.13, 7.36, 8.16, 9.15, 10.19, 11.10, 12.20
    Atterberg limits, 2.18
    Constant head permeability test, 3.20
    Critical hydraulic gradient, 3.24
    Effective stress, 3.23
    Estimating the coefficient of permeability, 3.21
    Equivalent horizontal permeability, 3.20
    Equivalent vertical permeability, 3.21
    Falling head permeability test, 3.20
    Flow net in isotropic soil, 3.25
    Foundation settlement in cohesive soil, 4.28
    Grain size distribution, 2.16
    Heaving, 3.26
    One-dimensional compression, 4.24
    Rock classification, 2.20
    Rock description, 2.20
    Seepage pressures, 3.24
    Soil classification, 2.18
    Time rate of consolidation, 4.30
    Vertical stresses due to surface loads, 4.25
    Well pumping test, 3.21-3.23
    Weight-volume relationships, 2.22

F

Falling head permeability test, 3.4
Field compaction equipment, effect of, 12.9
Field compaction of cohesionless soils, factors affecting, 12.3
Field density, 12.14
Field loading tests in axial compression, 9.4
Field tests for permeability, 3.8
Field vane shear test, 7.16
Field void ratio-pressure relationship, determining the, 4.7
Flexible retaining walls, 10.2
    Design of, 10.12
Flow net, drawing a, 3.17
Footings of finite lengths, 8.4
Footings on clay and plastic silt, 8.13
Footings on sand and nonplastic silt, 8.11
Footings with inclined load, 8.4
Foundation settlement, 4.18

G

General sources, 7.3
Geologic data sources, 7.2
Geophysical methods, engineering, 7.4
Gravity flow, 3.10
Gravity retaining walls, 10.1

H

Handheld devices, 7.18
Handheld sounding rod test, 7.23
Heaving, 3.19
Hydrometer test, 2.3

I

Identifying characteristics of soils, 2.10
Infinite slope analysis, 11.5
Initial relative density, 5.10
In Situ character, 2.12
In Situ testing, 7.14
Intact character, 2.12
Interpretation of pile load tests, 9.5
Isotropic soil, flow net in, 3.15

L

Laboratory consolidation test, 4.20
Laboratory oedometer test, 4.4
Laboratory relative density tests, 5.9
Laboratory shear tests, types of, 5.6
Lateral earth pressure, 10.2
Lateral loading of piles and piers, 9-12
Lateral loading tests, 9.4
Lift thickness, 12.19
Lift thickness and magnitude of effort,
    effect of, 12.4, 12.9
Liquid limit test, 2.5
Liquidity index, 2.5
Load carrying capacity of a single pier, 9.8-9-12
Load carrying capacity of a single pile, 9.8-9-12
Log of time fitting method, 4.22

M

Mixed grain soil, 6.12
Modifying compactibility with chemicals, 12.12
Modulus of subgrade reaction, 8.14
Mohr's circle of stress, 5.2
Mohr-coulomb failure criteria, 5.3
MSE retaining walls, design of, 10.10

O

One-dimensional compression, 4.2

P

Passive earth pressure, 10.8
Pavement design, 8.15
Penetration test data, analysis of, 7.23
Permeability, coefficient of, 3.1
    Laboratory tests for the coefficient of, 3.3
Permeability of rock, 3.7
Pile capacity from driving data, 9.14
Pile driving formulae, use of, 9.14
Pile load tests, interpretation of, 9.5
Pits, 7.6
Plate load test, 7.14, 8.13
Plastic limit test, 2.5
Point load, 4.11
Pore-water pressure, 5.8
    Negative, 5.11
Preconsolidation pressure, determining, 4.5
    Estimating, 4.9
Pre-existing information, sources of, 7.2
Pressure distribution with depth, 12.2
Proctor compaction test for cohesive soil, 12.13
Project records, 7.3

Q

Quality assurance definitions, 12.12

R

Rafts on sand and nonplastic silt, 8.12
Rankine's theory, 10.5
Relative density, 5.9
Remote imaging, 7.3
Residual soils, 6.4

Retaining structures, types of, 10.1
Rigid footings, stresses under, 4.15
Rigid retaining walls, design of, 10.9
Rock, definition of, 2.1
Rock classification, 2.12
Rock identification, 2.12
Rocks, igneous, 6.1
    Metamorphic, 6.1
    Sampling of, 7.10
    Sedimentary, 6.1

S

Samplers, 7.10
    Diamond-core barrel, 7.13
    Thick-wall split-barrel drive, 7.10
    Thin-wall tube, 7.11
    Vibrating tube, 7.11
Seepage line, 3.18
Seepage of water through soils, 3.15
Seepage pressure, 3.11, 3.13
Seismic refraction surveys, 7.4
Seismic slope stability analysis, 11.10
Selection of shear strength for design, 5.19
Sensitivity of clays to disturbance, 5.18
Settlement, 8.1
Calculation of, 4.9
Settlement limitations, 8.8
Settlement of pile groups, 9.13
Shallow footings on or near slope, 8.4
Shear failure, 8.2
    General, 8.2
    Local, 8.2
Modes, 8.2
Punching, 8.2
Shear strength of partially saturated soils, 5.17
Shear strength of soils, tests for, 5.5
Shear strength relationships, 5.12
Shrinkage limit test, 2.6
Slices, method of, 11.8

Sliding wedge analysis, 11.2
Slope failures, types of, 11.1
Slope stability charts, 11.6
Soil, definition of, 2.1
    Sampling of, 7.10
Soil classification systems, 2.6
    AASHTO highway, 2.10
    Unified, 2.7
    USDA soil taxonomy, 2.10
Soil compaction, specifying, 12.17
Soil deposits, natural, 6.4
Soil gradation, effect of, 12.3
Soil forming processes, 6.2
Soil structures, 6.11
Soil texture, 2.12
    Effect of, 12.8
Standard penetration test, 7.19
Static-cone penetration test, 7.20
Structural retaining walls, 10.2
Subsurface access methods, 7.6
Subsurface investigation, plan for a, 7.1, 7.32
Surcharge loading, 10.6
Square root of time fitting method, 4.23
Swelling index, estimating the, 4.9

T

Tension loading tests, 9.4
Test fill, example of, 12.7
Test loading and time relationship, 9.5
Testing, quality assurance, 12.12
Theoretical development, 4.18
Time rate of consolidation, 4.18
Transported soils, 6.5
Trenches, 7.6
Triaxial shear test, 5.5

U

Unconfined compression test of saturated
    clays, 5.16

Unconfined compression test of undisturbed
  cohesive sample, 7.17
Unconsolidated-undrained shear of saturated soils,
  5.16
Uniformly loaded circular area, 4.12
Uniformly loaded embankment, 4.13
Uniformly loaded line, 4.11
Uniformly loaded rectangular area, 4.12
Uniformly loaded strip, 4.12
Upward flow, 3.14

V

Vertical stresses due to surface loads, 4.10

W

Water content, effect of, 12.4
Water content test methods, 12.14
Weathering processes, 6.2
Weight-volume relationships, 2.15
Well pumping test, 3.8

Z

Zero air voids curve, 12.2

NOTES

NOTES

NOTES

NOTES

NOTES

NOTES

NOTES

NOTES

NOTES

NOTES

NOTES